LEAN MADE SIMPLE
Creating Pull & Flow

David Sykes

Published by Lulu.com

First published in the UK by Lulu.com

ISBN 978-1-71654-972-4

THE AUTHOR

David Sykes is by profession a chemical engineer but has spent most of his career in manufacturing and people management and latterly as a Business and Training Consultant.

In 1988 he was first introduced to lean principles in the guise of Just in Time when he was the manufacturing director of a company chosen as a JIT supplier to a major medical equipment supplier. In 1992 he founded JIT Services, a company specialising in reducing changeover times in manufacturing companies. Since then he has applied lean principles in the various and diverse businesses he managed.

After over 35 years in manufacturing, David founded Vanilla Training Solutions in 2005, a consultancy committed to helping organisations excel through the strategic use of the training process. He is the author of the books, Lean Made Simple, Creating Stability, Leadership, a Formula for Success and An Engineers Guide to Influencing and Persuading. He lives in Somerset with his wife Judith.

He can be contacted at david@theleadershipformula.uk

I dedicate this book to my darling wife Judith who has been my rock and my soulmate for the past 52 years. I also dedicate it to my six grandchildren, Elle, Lucy, Sofia, Miroslav, Thomas and Amelie who bring us so much joy. Each, different in their own way and yet so close, despite the distance between them. They are a credit to their parents and, if I am ever tempted to despair about the future of humanity, I have only to look at them to know it is in safe hands; six wonderful children. Hopefully, one day they may read my books and realise that Granddad didn't just tell tall stories.

CONTENTS

INTRODUCTION

In my earlier book, Lean Made Simple – Creating Stability, I contrasted driving on a Smart Motorway with the normal stop-starts of a conventional motorway - a simple example of lean principles in action. Our natural tendency to optimise our journey time ironically creates, when the motorway begins to reach capacity, the instability that leads to stationary traffic and frequent accidents, particularly rear-end shunts. The techniques employed, such as overhead speed restrictions and traffic lights on slip roads, apply the principles known as 'pull' and 'balanced flow' and we discuss these in this, the second book in the series.

If you are not familiar with Smart Motorways (I have sold many copies of my first book in the USA and driving to Yellowstone park in 2018 with its vast open interstates, I began wondering if the analogy had any relevance) they are a way to control traffic flows on a motorway and effectively increase capacity without having to build further lanes. I will explain the logic behind them later in the book.

If you haven't read my first book, don't worry, it may be that your organisation already has created stability and has sufficient procedures in place for you to be able to take the next step on our lean journey. I give a brief resume of the steps required to create stability in Appendix I. The importance of creating stability before introducing further lean techniques cannot be overstated. In the same way you would not build a house on shifting sands, it is pointless introducing techniques such as levelled scheduling in a business in a constant state of flux.

The key word to stress if we are to create a successful Lean Enterprise is - discipline. (Before any budding dominatrix reading this rushes out to scoop up every book on Lean, let me just say, it's not that kind of discipline!) Lean requires a disciplined approach from everyone and at every level in the organisation. The worst offenders, constantly frustrating lean techniques, overruling Kanban and balancing flow, for example, are often managers and supervisors. It is as if they have an inner compulsion to manage when management isn't required, to supervise when there is nothing to supervise. Instead of managing and leading their people, many managers want to 'play with the train set.' The reality is that the 'train' can happily run around the tracks without them thanks to the effort taken in designing (using lean techniques) the layout and the systems in the first place.

The key to changing this is to make sure managers and supervisors understand the principles that drive lean. Once these are embedded in their collective mind-sets they can then use their energies and skills to ensure that everyone else understands and follows the new ways of working. We will discover some effective techniques to overcome this when we discuss 'Leading4Lean' in Book 3. If you can't wait, I suggest you read my book, Leadership, a Formula for Success available from Lulu publications as well as Amazon and other online book sellers.

One additional benefit from implementing lean is that it can flatten the organisational structure. Introducing lean drives decisions and responsibility for the day-to-day production towards the work groups where they rightly belong.

So, what can you expect if you implement lean in your organisation? Typically for any type of operation, productivity will improve. Improvements I might add are not just marginal, we are not talking one or two percent, more in the order of a 50 to 100% increase. In other words, output per employee could double by adopting lean. We have already shown through our analysis of waste in Book One that there are many things we do in our organisation that are unnecessary. If you have adopted these techniques you have probably seen gains in productivity already by simply undertaking a waste analysis. Combine this with Value Stream Mapping which we consider later in this book, and you can really capitalise on the productivity benefits of lean.

Another benefit is that production lead times, the time between starting an activity and completing it, will reduce. Once again, the benefits are dramatic. Lead time reductions of up to 90% are common. A corresponding reduction in inventory is also to be expected. If it sounds too good to be true, for once, it actually is. No catches, no bullsh*t, trust me, I'm an engineer!

A substantial benefit we can expect with the implementation of lean is space savings. In my career in manufacturing and later consultancy, the most common complaint from managers, supervisors, team leaders and even operators is, 'we haven't enough space!' Lean offers a practical way of releasing space and up to 50% extra is not unusual.

It goes without saying that with all these benefits we can expect a dramatic improvement in both bottom-line profits and customer service. Profit, I often remind clients, is the difference between two, usually large, numbers. To improve profits we must increase one, income through sales and reduce the other, costs. The savings that lean generates must be real if we are to reduce costs. What do I mean by real? I am reminded of the story of the little boys who came home one evening and said to his father, "Dad, Dad, I saved a pound today!" When his father asked him how he said he had run home behind the bus and saved his fare. His dad looked surprised and then said, "Well tomorrow, son, run home behind a taxi and save five pounds!"

The savings lean generates are only real if they lead to reduced costs or increased sales. Some consultants will tell you they have given you 'opportunity' cost savings; time saved doing a job releases time to do other things. Until you have realised those savings, however, you are like the boy running home behind the taxi - releasing them only occurs through fewer shifts, less overtime, fewer people, less supervision or more volume throughput due to increased sales. We must never lose sight of this when we implement lean. Hopefully by being more productive, our cost reductions can lead to lower prices and higher sales which can mop up some of the excess manpower released.

A final benefit, not to be underestimated, is an improvement in employee morale as they become empowered and engaged at every level.

These benefits I might add do not come about by chance and are not immediate, only after a detailed analysis of the operation and the rigid application of lean techniques. The seven stages of problem solving we discussed in Book One are essential to this process and the above benefits will be delivered consistently only after the full implementation stage.

When I used to make this my sales pitch to potential clients, I would pause at this point and, in the fashion of Clint Eastwood in the film Dirty Harry, I looked them in the eye and said, "the only question is, are you brave enough?" (I wisely omitted the word 'punk' at the end of the question!)

Up until now, the lean techniques in Book One give us stability without much risk. In fact, many of the techniques have been around for decades and are simply, good manufacturing practice. Once we begin to introduce flow and then pull, however, the stakes become higher. We have to give up our old comfort blankets of stock and long production runs and really get to grips with solving our underlying processing problems. So, yes, there are risks, but thankfully others have gone before us and have shown us how to overcome them.

There are many hurdles along the way which we will encounter. Since lean often appears counter-intuitive, there is the temptation to override the systems, particularly Kanbans and go back to the old ways of working. Only by fully understanding the theory behind the changes will those involved in the process avoid these temptations. This is the purpose of the book you are about to read.

I remember one lean implementation project. One of the team members, the shift production manager, was convinced very early in the project that he knew what the outcome would be. He spent time on his own, away from the project team, and came up with a series of layout drawings he was convinced would be the 'final state' as it is known in Value Stream Mapping (VSM). I asked him to put them to one side and to keep an open mind until we had mapped the process completely and could analyse the data.

Needless to say when the final state was defined it looked nothing like his drawings. I explained that, although I guessed his drawings were wrong, the only way I knew what the final state would be was by strictly following the VSM methodology. This is the lesson I would ask you to remember. The techniques in Book One stand on their own but to implement lean effectively requires a detailed analysis of your operation. This is the purpose of this book, Lean made Simple – Creating Pull and Flow.

Who is this book written for? Primarily, for those who already know about lean, have introduced methods to create stability and are ready to introduce more powerful ways of reducing lead times and costs. (For those investigating lean for the first time, I strongly recommend my book, Lean Made Simple – Creating Stability published by Lulu and available from Amazon and all good on-line booksellers). It does not pretend to give you a comprehensive, blow-by-blow methodology required for a full lean implementation. The key word in the title is 'simple'. What it will do, however, is give you an insight into key elements of lean, in LAYMAN'S terms, which you can use to supercharge your organisation.

If you are a manager, supervisor or team leader considering a transition to a lean enterprise, not only do you need a broad understanding of lean, but a different approach to leadership. In the same way as driving on a Smart motorway instils in the users a different way of driving, lean systems require a different style of leadership. I call this style, Leading4Lean, a programme I developed in 2008 to assist those in leadership positions to manage more effectively within a Lean Enterprise. I cover this in detail in the last of the books. If you wish to know more about becoming a better leader, however, I recommend that you read my book, Leadership, a Formula for Success.

So, what's in it for you? If you have already introduced the techniques to create stability in your company and are ready to progress to a full lean enterprise, this book is written for you. Even if you are not ready to introduce Flow and Pull, you will still benefit from the module on setup time reduction. In this book, I offer you real-life examples where you can see how lean has been introduced into organisations as varied as carton manufacture, injection moulded containers and a semi-conductor FAB. If lean sounds as if it could be expensive, believe me, changes can be made at very little cost and, in the long run, are often self-financing. Whatever your reason, I guarantee that the knowledge you acquire from this book will help make your company more profitable and your job better and easier. Trust me, I'm an engineer!

David Sykes
September 2020

MODULE 1

CONTINUOUS FLOW

We live in a stop-start world. How else could it be? Our daily lives are broken into discreet segments from the moment we wake to the moment we go to bed again. When we work in an environment, therefore, where tasks are carried out in a similar fashion, flitting from one activity to the next, we see nothing unusual in this. In fact, we have a saying that captures its seductive attractiveness; **'variety is the spice of life'**. The problem with stop-start working is that it engenders waste in all the forms we identified in Book One. (For a resume see Appendix 1)

In manufacturing companies a concept was created to try and minimise the costs associated with stop-start working. It was called 'Economic Batch Quantity' and a formula was devised in 1913 by W Harris and R H Wilson to calculate exactly what might be the economic quantity to manufacture for any particular product. Clearly, it was thought that, in a volume production process, manufacturing only one of an item was undesirable particularly in terms of profitability. Set-up times relative to production times would mean idle machinery, and machinery in whatever form, is never cheap. On the other hand, manufacturing thousands of pieces, whilst ideal in terms of 'efficiency,' was clearly wasteful if demand was considerably lower thereby requiring massive amounts of finished stock-holding. EBQ therefore was the compromise and was endorsed by its supporters with scientific reverence.

Henry Ford, however, had other ideas. In the early days of motorcars, manufacturing these mechanical marvels was considered to be an art and they were lovingly produced by skilled craftsmen. Adam Smith in the Wealth of Nations in 1776 had already told us of the folly of generalisation; of 'trying to do everything' ourselves. His theories pushed the world towards individual specialisation as the way to economic success. Ford took up the message and created the first automobile assembly line where specialist 'wheel-nut turners' could practice their skills without ever having to worry about the intricacies of seat-fitting or headlight alignment. In this way continuous production of motor cars was developed with no need for the calculation of economic batch quantities. The assembly line turned out automobiles quicker and cheaper than his rivals, the only proviso as we all remember, they had to be black!

Henry Ford provides us with an obvious example of Continuous Flow. It is also known as one-piece flow and the logic is very simple. Make something in one go. No stopping and starting, no sub-assemblies, no work-in-progress (WIP), no Economic Batch Quantities. As the man used to say on the TV show, Mastermind; "I've started so I'll finish."

All very well, I hear you say, if you are assembling cars or making chemicals, but my production processes do not lend themselves to this type of continuous flow. (Neither did automobile manufacture before someone worked out how to do it!) Quite often the reasons for this are simple. When setting up production processes we have a tendency to balance **capacities** in our operation rather than balancing **flows**. What do I mean by this? Think of a product; let us call it a 'widget gadget' consisting of three components, A, B and C. Component A is manufactured on an old forming press which can handle 1,000 components an hour. It then goes to the 'grunging' machine which 'grunges' component A with the main body B to produce a semi-finished 'gadget'. The company has bought a state-of-the-art machine that can spew out 3,000 components an hour. It is completed on the 'finishing' machine which adds the final component C and can produce 1,500 completed 'widget gadgets' an hour. The production planning department has scheduled 24,000 units a day. I'm sure you're ahead of me...

The forming press needs to run three shifts a day, building a stockpile to feed...

The 'grunging' machine which runs one shift a day building stocks to feed...

The finishing machine which runs two shifts a day to complete the quantity required.

Overall the total **capacity** of each stage of manufacture is 24,000 units a day, whilst the **flow** varies from 1,000 units per hour to 3,000 units per hour.

If this sounds familiar, that's fine, this is how manufacturing has developed in many operations and the purpose of this module is to explain the process of achieving continuous flow. 'Flow' is essential if we are to take advantage of the powerful tools that lean affords us.

So, what are the benefits of one-piece flow?

Quality. If you are familiar with producing in batches I am sure you have come across the problem of starting to process the stock delivered to the line only to find there are defects. Instead of producing we end up either; sorting the stock, rejecting it or processing it and then placing it on hold for someone to sort out the mess! In-line production avoids the problem as every operator is also an inspector and can detect faults and more importantly, ensure they are fixed as they arise.

Inventory: By converting raw materials and components into finished product in one operation the need for Work In Progress (WIP) is mostly eliminated. (We may need to store one or more units, which we will discuss when we consider Kanbans) Not only does this free up space, it frees up capital.

Space: The floor space released through needing less inventory is only one of the areas where space can be freed. By reorganising equipment to create flow we also release floor space, something which is often overlooked. I remember working as a consultant in a semi-conductor factory and looking over the final operations area as we first toured the FAB (the name for a semi-conductor plant). It looked like the London branch of PC world with long benches everywhere crammed full of test equipment and monitors. The poor operators moved frantically around the vast area looking for signals to tell them that the test cycle was completed on an individual piece of test kit. Once we reorganised the area into production cells, we could have held the works Christmas party in the space we cleared. Space by the way which, because of the clean room requirements, cost £8,000 per square foot to build!

Flexibility: Organising resources differently can give us the flexibility to shorten lead times and the ability to respond quickly to customer demand. Don't get hung up on the idea that continuous flow results in every factory or operation becoming an assembly line like the Ford example I gave. All types of processing operations whether garment factories, bottling plants or decorative cake making can move to one-piece manufacture once the principles are understood and applied.

Productivity: Chasing high levels of productivity is often the Holy Grail of manufacturing operations with machine utilisation and labour efficiency driving behaviours across the plant. I have seen managers become obsessed with achieving 'record' productivity, sometimes by cherry-picking the easy jobs from the production plan, other times over-running an order because the line was running so well! But how much real productivity are we achieving when we are creating waste in the form of over-production, defective parts, surplus inventory and the like? In one-piece flow there is very little activity which does not add value to the product. When we observe the people involved in the new operation, it becomes clear who is busy and who is not.

I once was the manufacturing director of a business comprising of two factories. One was managed by an ex-German Prisoner of War who I shall call Fritz who some described as a 'slave driver.' The other was managed by a mild-mannered man who was comfortable with figures but avoided people as much as possible. He spent his days, not managing the people but re-planning the production because, whilst he was cocooned in his office, nothing in the factory ever happened the way he had planned it!

In Fritz's factory he was 'here, there and everywhere' and everyone was always busy; in fact no one was allowed not to be. The factory had little automation and the operators sat at their workplace, injection moulding machines, which dropped perhaps 8 components every 12 to 16 seconds into a collection tray beneath. In Fritz's factory, the operator would pick up the mouldings and trim off the 'flash' or surplus material before layer-packing them into a cardboard box. Impressive you might have thought whilst watching it, except, on closer inspection, there wasn't any flash to remove! If there had been the engineering manager would have removed the tool and promptly repaired it. This fantasy 'flash removal' continued until the start of the next cycle.

At the other factory, it was a completely different scenario. Operators sitting on similar machines sat waiting patiently for the components to drop which they then packed into a box before waiting, patiently, for the machine to offload again.

Fritz boasted that no-one ever saw anybody not working in his factory and claimed it to be the more productive of the two. Since the productivity was calculated using the standard labour hours recovery we introduced in Book One, actually they were both approximately the same. Fritz was making the mistake many of us make; *he was confusing activity with action.* In time, the other factory, under different management, became the most productive. I was able to re-organise the operators and, with the expense of a few packing tables, use the spare time of the operators to perform additional tasks, reducing the need for assembly work elsewhere. In the long term I closed the 'busiest' factory and transferred the work to the other. By then, it had the highest 'actual' rather than imaginary productivity of the two.

Safety: One-piece flow drives us to manufacture smaller batches and less stock which leads to fewer movements around the workplace. The simpler layout that arises out of continuous production results in a safer workplace.

Morale: People doing more value-added work generally have a higher level of job satisfaction, motivation and morale. Creating Flow allows us to develop self-directed work groups of the kind we discussed in Book One.

Our starting point to achieve Flow is to introduce the concept of 'Takt' time, the fundamental driver of a lean enterprise.

TAKT TIME

The first question we must ask ourselves in any manufacturing operation is how do we decide at what **rate** must we produce? The quantity to be produced is likely stated on the customers purchase order. If we are selling our own product from stock or are contracted to hold customer stock, then we have licence to decide the production quantity. Since production is closely linked to material procurement, companies have gravitated to using scheduling and forecasting tools to answer this fundamental question. MRP, MRPII and SAP spring to mind.

However, customer demand can change rapidly and markets change constantly so scheduling and forecasting systems are by their nature fundamentally flawed. This is where Takt time can offer us clarity. Takt is a German word and it describes the baton wielded by the conductor of an orchestra. It is the 'drumbeat' or pulse that that every member of the orchestra is aligned to.

Takt is the principal that all activity within a business is synchronised by a pulse, not picked out arbitrarily by the planning or production or engineering departments, but set by the **customer demand**.

Takt time links the pace of sales with production, mandating that a production system is flexible and linked to actual customer demand. It is used to ensure that all operations and processes are synchronised to the same drum beat. Once we have established the Takt Time all operations in our production chain can then be 'tied together' by introducing other elements of lean. So, you've got my attention I hear you say, how do we calculate Takt Time?

Takt time relates the customer demand to the time available with the formula:

$$\text{Takt} = \frac{\text{Production Time Available}}{\text{Customer Demand}}$$

The available time has to take into account such factors as:
- Lunch and tea breaks
- Team briefing times
- TPM breaks
- Clean down time

Note the available time does not take into account any lost time due to breakdowns and other unforeseen events; we allow for this at another stage.

In the example I gave it is clear that in using 8 hours I did not allow for the unavailable time in my calculation. Assuming a 10 minute tea break and 30 minute lunch break together with a 5 minute clean down time at the end of the shift, the available time for an 8-hour shift would be:

$$(8 \times 60) - 10 - 30 - 5 = 435 \text{ minutes}$$

If in our hypothetical example, despite the production capacity which was planned, the actual customer demand is 100,000 units per week. This calculates as 100,000/5 = 20,000 per day. The Takt time calculation is therefore:

Takt Time = 435/20,000 = 0.02175 minutes or 1.305 seconds for single shift working.

For working around the clock the available time is 1,305 minutes giving a Takt Time of 0.06525 minutes or 3.915 seconds. This is the market demand that our manufacturing processes have to fulfil.

To state the obvious, the longer time we have available, the slower we can make the product.

TAKT, CYCLE, TARGET CYCLE AND LEAD TIME

Let us introduce some other terms that will help us understand how we use Takt Time as the drumbeat for our operation. Remember:

Takt Time = the pace at which the customer requires products

We next need to define the cycle time we intend to produce to, this is called the:

Target Cycle Time = the pace at which we will produce to ensure we meet the customer requirements

We usually set the target cycle time faster than the Takt time in order that we can stop production in the event of a problem without immediately effecting the customer delivery.

The machinery or process we use has its own cycle time, independent of Takt or Target CT, this is known as:

Cycle Time = the time at which a process cycles

Lastly, whether we are using one or multiple processes, the total time to produce a product is called:

Lead Time = the total production lead time from product start to finish

(This is not, please note, the 'lead time' we might give to the customer which takes into account many other factors and is generally the time between receiving an order and delivering it.)

To show this visually:

Takt, Cycle, Target Cycle and Lead Time

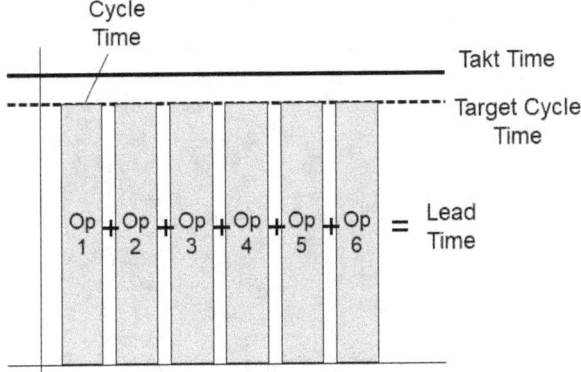

As an old production hand, when I was first introduced to this concept I found it difficult to grasp. When thinking of machinery, I grew up talking of units produced per hour or for ultra-fast machinery, parts per minute. Turning this on its head we now need to think of the time required to manufacture one unit. The advantages of this will become clear as we progress through the book.

Let us look at a practical example of creating flow without, at this stage, worrying too much about detailed timings. Traditionally, machine shop operations will schedule by the process used. For instance all the lathe work will be done in one area, the milling in another, grinding in another and milling somewhere else.

Semi-finished product transfers between workstations and is stored in the area local to the process. Often work will return to a previous operation and if we were to track the progress of one item during its manufacturing, it would almost look like a spider trying to form a web. If we were to create a diagram tracking the activity, we would see there was no evident organisation or control. Let us consider the machining of one particular product. In simplified form, the material flow might look like this:

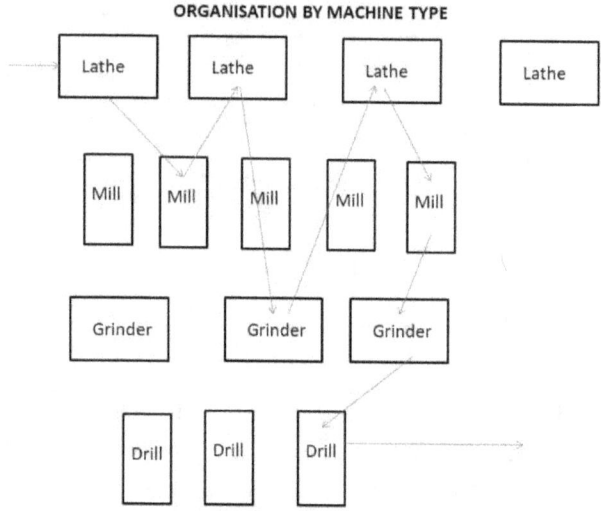

ORGANISATION BY MACHINE TYPE

Of course, there is more to this layout than is shown on the diagram. Each machining station has a bin of materials awaiting work and a bin of finished material awaiting collection. These bins have to be removed to the next processing station before it can process it. This station will also have its own allocated space for WIP. In each area there may be one or two operators to load, operate and offload the machine.

When we create flow, we look at moving the individual work centres required into a line, one next to the other. An ideal layout for this is a U-shape as this reduces the length of the line and allows operators to access both sides at the same time. This also facilitates loading and unloading the production line from the same end. Commonly called 'cells' a one-piece flow layout for this component is shown below:

U-SHAPED ONE-PIECE FLOW CELL

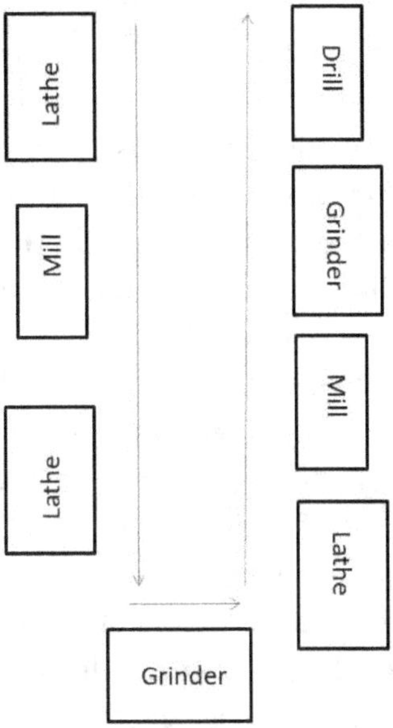

It looks so simple, doesn't it? But easy? Not so. There are many hurdles which must be overcome to introduce flow into a manufacturing operation. The first obvious one is that if any of the machines in the cell goes down, the whole process stops. In the original setup, each work station operates with its own buffer stock so if one machine goes down, the others can continue producing. Production people have it in their DNA that machines must be kept running so any reorganisation that puts at risk the whole output of the facility is viewed by those operating it with extreme distress! This is one of the main reasons why we need stability in an organisation before implementing flow. Total Planned Maintenance and problem solving techniques need to be in place before we can seriously implement flow.

I have seen operators, frustrated that the work station ahead of them has gone down, continue producing and stacking semi-finished product first onto the floor and then, when that becomes full, on top of their own machines! This, ironically, is actually counter-productive and introduces further disruption to the line. When setting up the cell, the designer will have balanced the throughputs of each of the resources (how we do this we will come to later) so that, when the work-station ahead is back on line, it does not have the capacity nor does the operator have the time to process all the unexpected and unnecessary WIP that has been created.

Another of the hurdles encountered is backtracking from senior management when difficulties arise. With a troublesome machine in the line, the temptation is to start producing sub-assemblies which can be taken to another machine to be completed. (I confess to have done this myself but only as a temporary measure whilst the root-cause of the machine problems is established and resolved). If we take the soft option, the pressure to resolve the problem of unreliable equipment disappears. The solution may lie in an additional investment which managers may shy away from. Our role as managers, however, as we will discuss in Book Three is to **make the right things happen through people**, not come up with excuses, however reasonable, as to why it can't be done.

We have to think of Continuous Production as a chain in which each link is joined to the one on either side. When we try to 'push' work through it, as the operator did when trying to build WIP when the next link had failed, it doesn't work. Chains only work when we put them under tension; they are designed for pulling not pushing. How we create the 'pull' to ensure work flows smoothly to the customer to the Takt Time we have identified is the essence of this book.

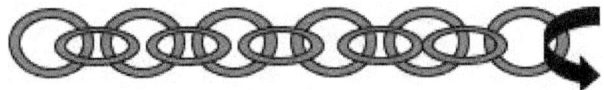

Customer

Before we discuss how we create this 'pull' and all the elements associated with it, however, we must first understand how we can balance the one-piece flow line to optimise machine and labour utilisation.

LINE BALANCING

So here's the dilemma; we know what we want, we need to move our machines into a line or a 'U'-shaped cell to create flow. However, if we think about our first example, moving these three machines into a line, creating one-piece-flow or the 'chain' we compared it to, will mean that either, with one-shift working the line will be producing at the rate of the slowest machine and failing to meet the Takt Time or, with three-shift working, two of the machines will be doing nothing for a large proportion of the time. (I know, you've already worked this out for yourself).

We have two issues to resolve: balancing out the machine usage and balancing out the labour usage – this is the subject of the next Module.

MODULE 2

ACTIVITY BALANCING

As we start our program to introduce Flow and balance the activities of both the work centres and the operators, let us consider what must be in place before we can proceed.

Equipment Reliability

We spent considerable time in Book One looking at ways to improve the reliability of equipment. These included problem-solving techniques using the 7 tools of quality, root cause analysis and TPM. To explain the importance of maintaining high levels of equipment availability, let us go back to the machine shop we introduced earlier. If each of the resources in the original layout, organised by machine type, had an uptime of 90%, (in other words 10% of the available time the machine was broken) the buffer stock before each workstation would mostly be set by scheduling but at least high enough to cover the unpredictable times that the previous machine is down.

However, once these machines are placed in a line to create one-piece flow, the consequence of the loss of uptime becomes critical. Why? It's simple mathematics. Eight machines, each with 90% uptime are now in a line and feeding each other directly. The overall uptime of the cell will be:

$$0.9 \times 0.9 \times 0.9 \times 0.9 \times 0.9 \times 0.9 \times 0.9 \times 0.9 = 0.43$$

Uptimes of this level, 43% are critical to both customer lead time and cost of production and rightly unacceptable. The solution to avoiding this lies in adding a few pieces of WIP between operations in certain locations to increase the overall cell uptime to 90%. In some operations, this is as simple one piece before each workstation to create a buffer against interrupting flow. Long term, of course, the causes of the breakdowns need to be identified and resolved. (When we discuss pull and Kanbans, this will be more obvious.)

Consistent Capability.

This was the primary objective of 'creating stability' in Book One. Quality should not be hit and miss, it should be built into the process. We will look at capability studies later in the book. Methods of working should not be 'made up as you go along' but SOPs (Standard Operating Procedures) should be in place to ensure consistency. People, equipment and materials should always be available before production commences, failure to do this is one of the main reasons implementing flow is unsuccessful.

Balanced Cycle Times

Operation cycle times must be balance to the Takt time and the target cycle time. Uneven work times will create waste in the form of waiting time or overproduction. This is what we are about to discuss.

So, what do we mean by activity balancing?

- Everyone is doing the same amount of work
- Doing the same amount of work to customer requirement
- Variation is 'smoothed'
- No one is overburdened
- No one is waiting
- Everyone is working together in a BALANCED fashion

Let us consider a simple example to illustrate the idea.

Line Balance : Simple Example

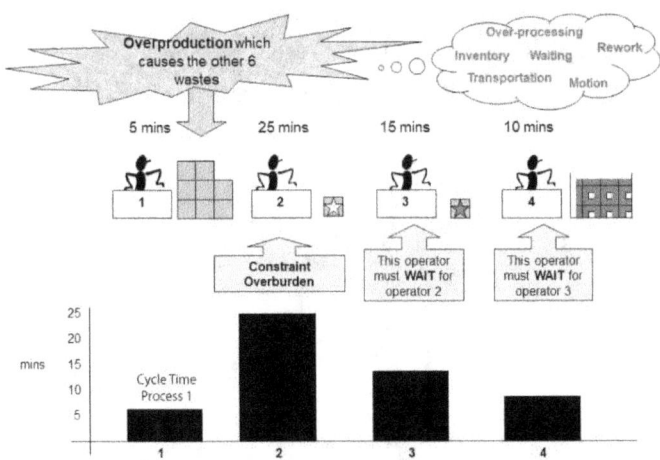

Here we have four processes, each with different cycle times. As we can see, operation 1 has the shortest time and operator 2 has the longest. Therefore, operator 2 is the CONSTRAINT or bottleneck. This means that operator 2 will have to do more work than the others - he is clearly overburdened. Since operator 1 has the shortest processing time it, he or she either sits and WAITS for operator 2 or OVERPRODUCES to fill their time. This creates the INVENTORY that you can see after stage 1.

The other downstream operations have to WAIT for Operator 2 also as their cycle times are lower. If Operator 1 does overproduce, then this tends to create the other WASTES in the system. We identified the wastes in Book One but as a reminder they fit into 7 categories.....

Overproduction, Waiting, Transportation, Over-processing, Inventory, Scrap/Rework, Motion.

If we were to re-allocate some of the work from operator 2 onto operator 1 then we can see how this can make the line more BALANCED.

This reduces VARIATION in the process, minimizes the 7 wastes, avoids overburdening the operators and promotes one-piece-flow.

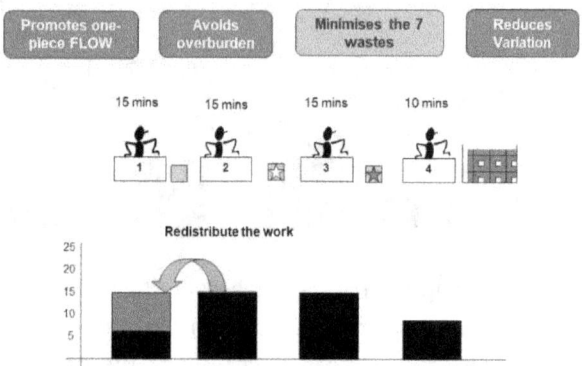

Once again, all of these attributes help achieve a stable work environment in which variations can be detected quickly and dealt with. The next processes down the line or if this is dispatch, the customer, receives a consistent supply of product.

Due to the low level of inventory in the system, if the customer makes major changes to the order, the system allows us to be flexible and we do not end up having to scrap large amounts of stock.

Rather than working in batches as previously, this system also encourages us to change over more frequently. At first this may be an issue as the process may not be set up for frequent changeovers. However, this is where changeover reduction activities can be used which we will discuss later in the book.

Before we can begin line balancing, we must establish the Takt time as we discussed earlier. This gives us the drum beat of customer demand.

If we know how many products we need to produce at regular intervals then we can balance production to meet this, ensuring, at the same time, that we always meet our customer requirements.

It also enables us to calculate the required level of manning for the process.

We discussed the need to standardize operations in Book One and already know it is a pre-requisite before we can introduce flow. What we mean by this is that by STANDARDISING the operations we have a base line on how the job should be completed. This enables better management of the process and problems or process variations can be determined more effectively. This is vitally important as one-piece-flow demands that we are able to resolve issues quickly before it affects deliveries to the customer.

We are now going to look at the process of line balancing. We will look at a simple assembly operation involving three workstations to illustrate the methodology. To do this we need to introduce other elements of standardisation, the **standard work chart** and the **standard work combination table.**

Standard formats, as we have seen, are popular tools for detailing the way that process should be run or managed.

Here is an example of a 'Standard work layout' document which would be used to detail:

- How the cell should be laid out (position of machines/ people / stock locations
- How the material should flow through the area.
- It also highlights Health and Safety information.

This document should be signed off by the team leaders of the area and the local area manager. It should be visibly displayed in the area for all to see (along with other standard documents or on a visual management board.)

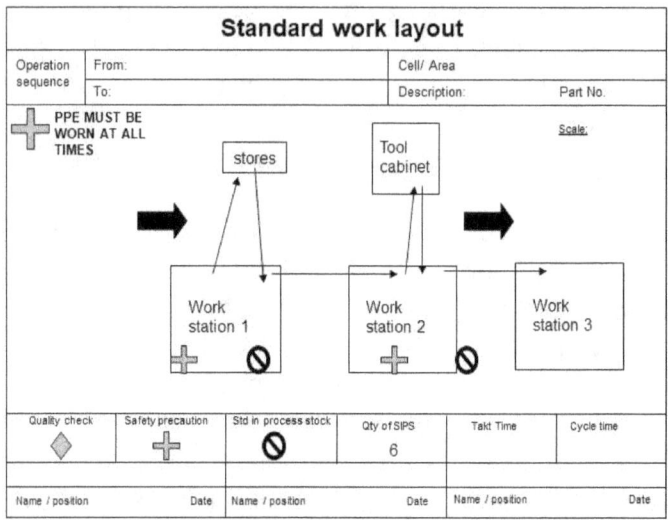

Standard work layout

| Operation sequence | From: | | Cell/ Area | |
| | To: | | Description: | Part No. |

We now need to capture the details of our operation on a 'STANDARD WORK COMBINATION TABLE (SWCT). This document is used for detailing each of the process elements involved. It is the record of how the job is actually carried out and can be used at a later stage for balancing the work between operators.

The Standard Work Combination Table combines human movement and machine movement based on Takt time and is used as a tool to determine the range of work and work sequence for which a team member is responsible.

Human work and Machine work

The key notion for the elimination of waste and the effective combination of work on the shop floor is the separation of machine work and human work.

When we observe the work in which operators handle machinery, then that work can be classified into either machine or human work. Understanding the separation of human and machine work is the basis for understanding the interface between these two elements. If operators are merely observing the machine working then this is the waste of "Waiting" and should be eliminated.

Human Work refers to work that cannot be completed without human effort. Examples of this are, picking up materials, placing materials onto a machine, operating the controls of a machine etc.

Machine Work refers to work or incidental work that equipment, which has been started by human hand, performs an operation automatically. For example, milling, auto riveting / bolting, auto inspection

Standard Symbols

The four basic symbols used in Standard work combination tables are:

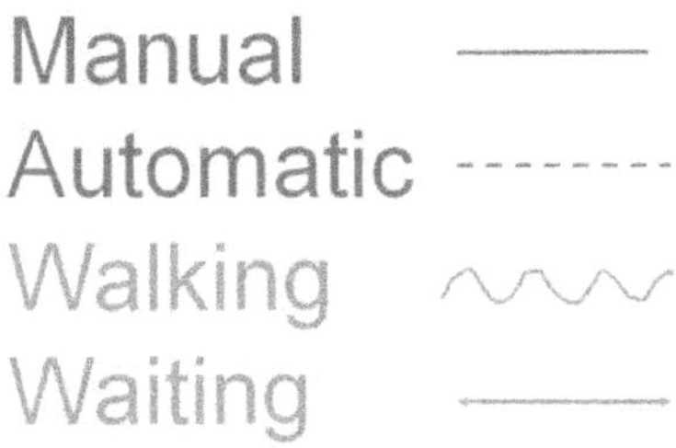

Manual   ———
Automatic -------
Walking   ∿∿∿
Waiting   ———

We display the results of our exercise as follows:

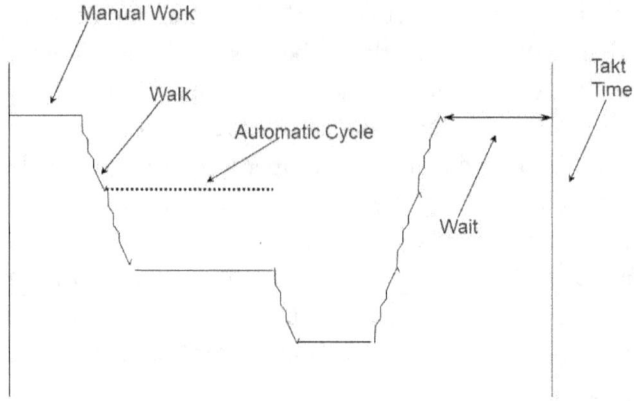

Manual Work

Walk

Automatic Cycle

Takt
Time

Wait

Of course, this is difficult to work with as it does not show timings; therefore we transfer this to our chart:

STANDARDISED WORK COMBINATION TABLE

Issue date 30/09/99

Part Number / Name	123ABC / Widget LH
Process	Machine and sub assembly

Takt Time: 162	Qty / shift: 169	Group Leader
Cycle time	Available time: 480'	Supervisor
	Department: Machine	Prod/Engineer

Manual ——
Automatic ·········
Walking ∿∿∿
Waiting ↕

OPERATING TIME IN MINUTES
5 10 15 20 25 30 35 40 45 50 55 60 65 70 75 80 85 90 95 100 105 110 115 120 125 130 135 140 145 150 155 160

WORK SEQ	OPERATION	TIME MAN	AUTO	WALK
1	Select part A	2		3
2	Set into M/C 1	10		
3	Start machine	1	84	2
4	Select part B	2		2
5	Set into jig	9		3
6	Select part C	2		3
7	Fit C to B	15		3
8	Set C/B to m/c 2	11		
9	Start m/c	1	41	2
10	Remove A from m/c	15		3
11	Set A to jig	7		
12	Remove C/B from m/c	5		3
13	Screw C/B to A	15		
14	Check torque	8		2
15	Put in finished bin	2		5
	TOTALS	95	125	34

Operator 129
Wait time 33

We are now ready to begin the actual method of line balancing. The first thing we need to do is to calculate the TAKT time and record the current state of our production process. As an example, let us consider the manufacturing process for the following product:

Calculate TAKT

Customer demand = 19 units a month

Time available = 20 days a month

$$TAKT = \frac{\text{Available time}}{\text{Customer demand}}$$

$$TAKT = \frac{20 \text{ days}}{19 \text{ units}} \quad (\text{x 24 hrs in a day})$$

Current State	
Ops	3
TAKT	25hrs
Total work content	
Line Balance Ratio	
Line Balance Efficiency	

TAKT = 25 hrs

As yet, we only know the number of operators currently operating the process and the calculated Takt time. (For the purpose of this example, we will assume the target cycle time is the same as the Takt time.)

The first step in our line balancing exercise is to capture the current method of working. We do this by timing the process. The best way of achieving this is to use a video camera. *Why video?*

- It can visually record activity which can be reviewed by others
- It is an accurate method of recording
- It is irrefutable and unambiguous
- It is a digital way of establishing the methods in place

It is essential to involve the line operators in the process. There is a phenomenon known as the Hawthorne effect where a person's performance will automatically improve if they are being observed, particularly if we take an interest in their work. By involving the operators, after a while they will take the camera for granted and eventually forget it is there. (Videoing a process is actually more difficult than you think. Whilst we must concentrate on capturing the process rather than the operator, sometimes capturing the operator may be unavoidable; in these cases it may be prudent not to capture his or her mugshot!!) The performance we record once the operator feels comfortable with being filmed will then accurately reflect reality.

Once we have a representative sample of the process, the next step is to sit down with the operators and review it together. The purpose of the exercise is to break down the elements of the work and record a time for each one. Having done this, we must then decide which of the elements is value-added and which is non-value added. Waste and NVA can be identified more easily by discussing the video together and allows the team to identify things that they previously did not realise occurred.

The findings of our exercise are shown next:

The operators cycle is broken down into elements
These elements are put into three main categories, these being :

1. Working (man or machine)
2. Walking
3. Waiting

Calculate Takt time

Time the process

Break down the work elements

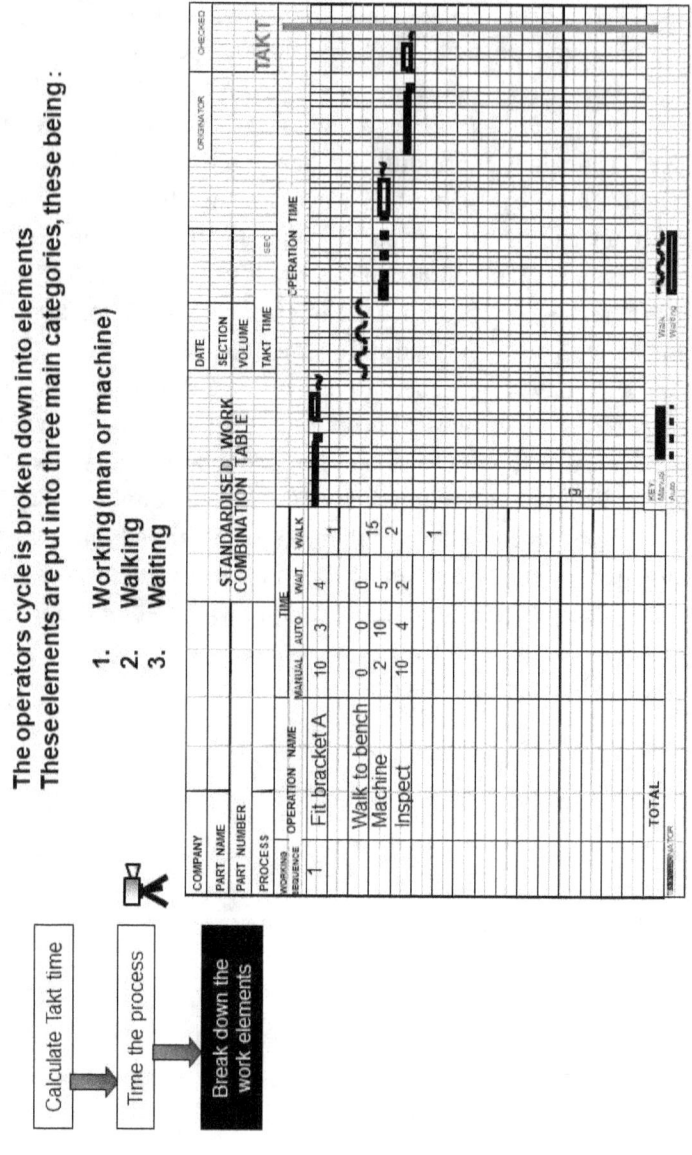

For each element of the process, we capture the part of the operation that is manual (for example loading a machine), those that are automatic and the time spent waiting and walking. This allows us to identify the total work content and the value-added time. We can then drawer the current state line balance, as shown below:

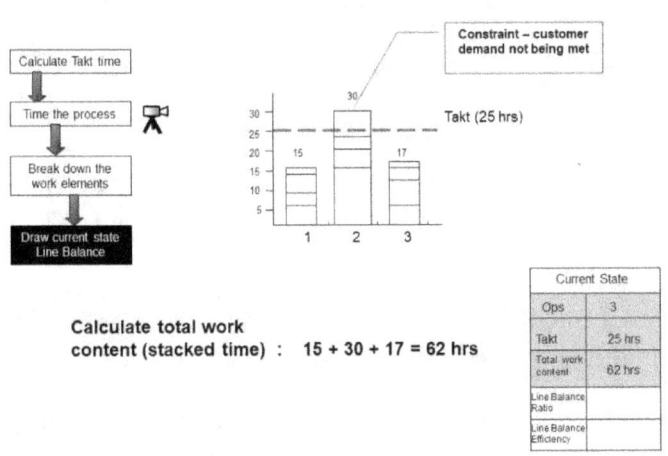

The next stage requires a large sheet of brown paper and some Post-it notes. Using the Post-it notes to represent each individual element of the process, we stack them on top of each other to identify the current cycle time of the resource. Once we have done this for all resources, we can add a piece of string to represent the Takt time, as shown in the diagram:

The target manpower can now be calculated by adding up the total work content of the job and dividing it by the TAKT.

This gives us something to aim for as it will tell us how the process could (in terms of the manpower required) run without any efficiency losses in a perfectly balanced state.

If the figure comes out as 2.48 (as in the example) then we will have to round the figure up to 3 and accept that there will be some spare capacity in the cell. If however, the manpower requirement comes out as 2.02 or 2.1 then we can justify only using 2 on the basis that the 0.1 can be reduced through a continuous improvement exercise.

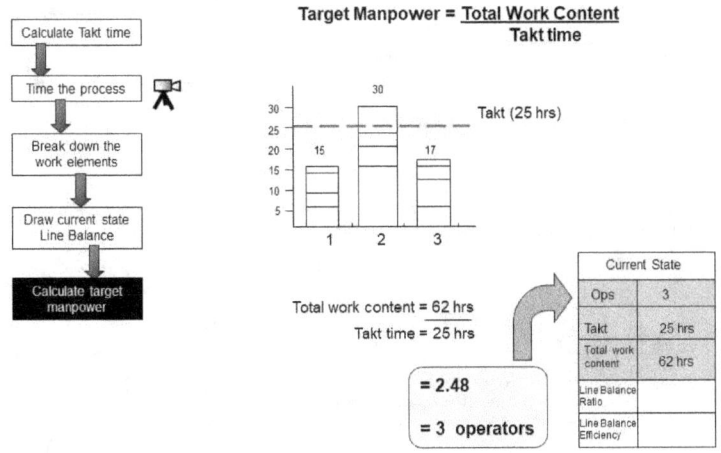

Remember we are still looking at the current state. This allows us to calculate two factors; the Line Balance Ratio and the Line Balance Efficiency. These are defines as:

$$\text{Line Balance Ratio} = \frac{\text{Total Work Content}}{\text{No of Stations x Longest Work Content}}$$

$$\text{Line Balance Efficiency} = \frac{\text{Total Work Content}}{\text{Target Manpower x Takt}}$$

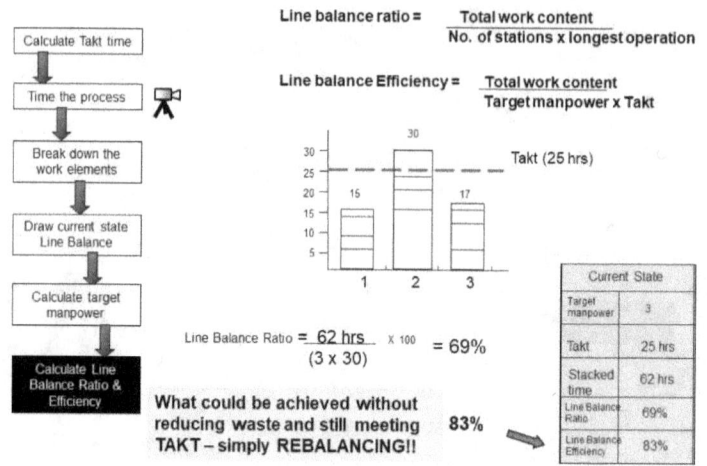

Calculate Takt time

Time the process

Break down the work elements

Draw current state Line Balance

Calculate target manpower

Calculate Line Balance Ratio & Efficiency

$$\text{Line balance ratio} = \frac{\text{Total work content}}{\text{No. of stations x longest operation}}$$

$$\text{Line balance Efficiency} = \frac{\text{Total work content}}{\text{Target manpower x Takt}}$$

Takt (25 hrs)

$$\text{Line Balance Ratio} = \frac{62 \text{ hrs}}{(3 \times 30)} \times 100 = 69\%$$

What could be achieved without reducing waste and still meeting TAKT – simply REBALANCING!!   **83%**

Current State	
Target manpower	3
Takt	25 hrs
Stacked time	62 hrs
Line Balance Ratio	69%
Line Balance Efficiency	83%

Already this tells us that even without making any improvements in the process, the customer demand can be met and a efficiency of 83% can be achieved. Clearly, in our analysis we have identified a high proportion of non-value added time which is ripe for reducing. The Flow Chart for how this is achieved is shown below:

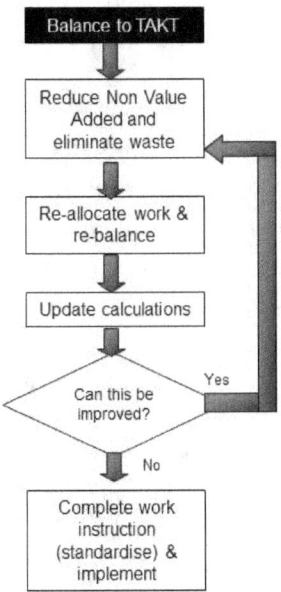

The first step is to identify the elements of the work that exceeds the Takt time. This shows us which work we need to redistribute.

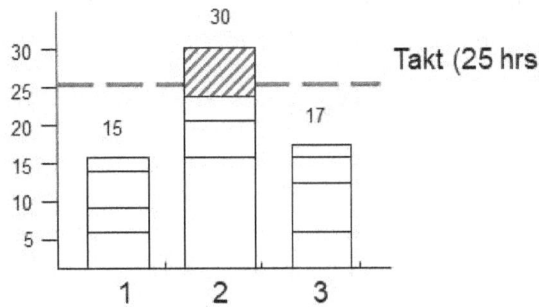

This is usually done by moving the Post-it notes around on the brown paper after discussion with the team leader and the operators. Referring to the Work Combination Table, look at where work can be re-allocated. In doing this, we need to identify the value added and non-value added work in three categories:

Value Adding:

- Any process that changes the nature, shape or characteristics of the product, in line with customer requirements e.g. machining, assembly. For example, a machine cutting metal is adding value, moving the part into position is not.

Non-Value Adding:

- Although non-value adding, some operations will be unavoidable with current technology or methods. It is defined as any work carried out that does not increase product value; for example inspection, part movement, tool changing, maintenance

Waste

- All other meaningless, non-essential activities that do not add value to the product that we can eliminate immediately e.g. looking for tools, waiting time.

We can now review the standard work combination chart and, using coloured marker pens, highlight in green the value-added work and in red, the non-value added:

COMPANY						DATE				ORIGINATOR		CHECKED
PART NAME				STANDARDISED WORK		SECTION						
PART NUMBER				COMBINATION TABLE		VOLUME				TAKT		
PROCESS						TAKT TIME	sec					

WORKING SEQUENCE	OPERATION NAME	TIME				OPERATION TIME
		MANUAL	AUTO	WAIT	WALK	
1	Fit bracket A	10	3	4	1	
	Walk to bench	0	0		16	
	Machine	2	10	5	2	
	Inspect	10	4	2	1	
	TOTAL					

KEY: Manual ▬▬ Auto ■■■ Walk ∿∿∿ Work / Waiting ▬▬▬

We can now go back to our representation of the current state and highlight the two categories:

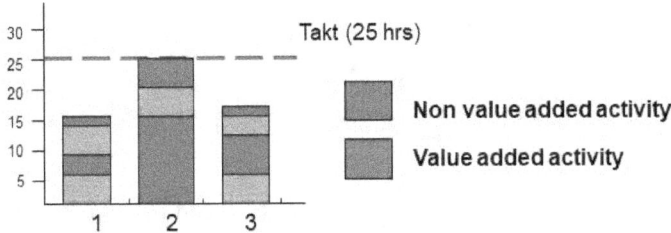

Takt (25 hrs)

Non value added activity

Value added activity

Our task now is clear; we need to remove as much of the non-value activity as possible and redistribute the value-added activities. We use the tools we already identified in Book One, including process mapping and 5S together with the changeover reduction techniques which we will discuss later in the book. The result will often be quite surprising:

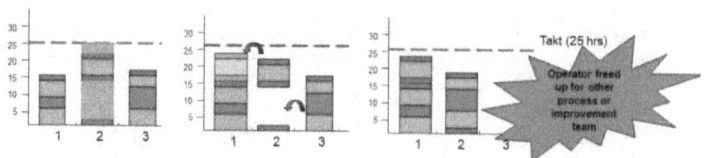

In this case, we see that we actually only need two operators to work the revised layout. In summary:

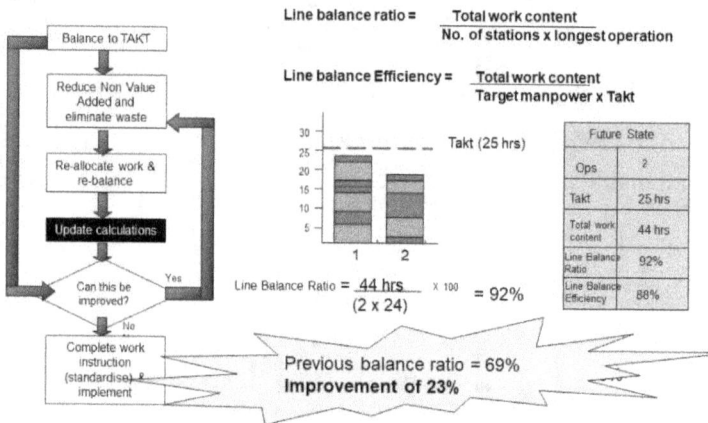

Line balance ratio = $\dfrac{\text{Total work content}}{\text{No. of stations x longest operation}}$

Line balance Efficiency = $\dfrac{\text{Total work content}}{\text{Target manpower x Takt}}$

Line Balance Ratio = $\dfrac{44 \text{ hrs}}{(2 \times 24)}$ X 100 = 92%

Previous balance ratio = 69%
Improvement of 23%

Future State	
Ops	2
Takt	25 hrs
Total work content	44 hrs
Line Balance Ratio	92%
Line Balance Efficiency	88%

A very impressive result! We have reduced the work content from 62 to 44 hours, reduced the number of operators required from three to two and rebalanced the production line to the Takt time. We need to capture the new layout and the new methods of working using SOPs.

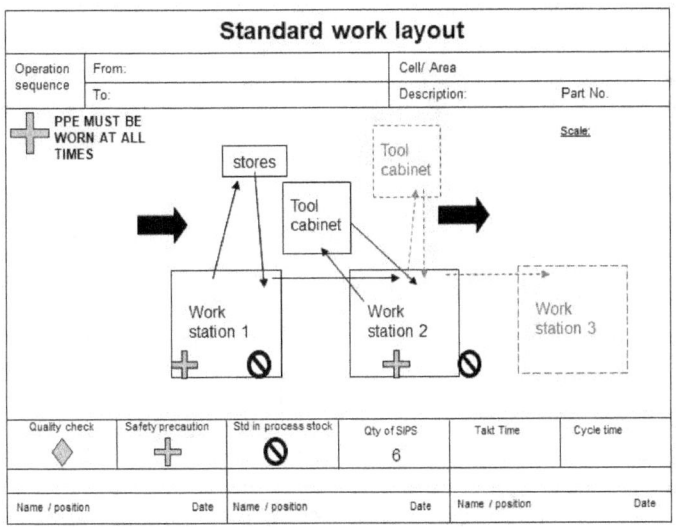

Standard work layout						
Operation sequence	From:			Cell/ Area		
	To:			Description:		Part No.

PPE MUST BE WORN AT ALL TIMES

stores

Tool cabinet

Tool cabinet

Work station 1

Work station 2

Work station 3

Scale:

Quality check	Safety precaution	Std in process stock	Qty of SIPS	Takt Time	Cycle time
			6		
Name / position	Date	Name / position	Date	Name / position	Date

Of course, we can further improve the process by using CI techniques to compress cycle time, one of which, setup time reduction or SMED we will discuss in the next module.

MODULE 3

SETUP TIME REDUCTION (SMED)

In 1988 I was the manufacturing director of a company supplying injection moulded packaging to a major UK medical equipment manufacturer. Life was good. We had a supply agreement that guaranteed a fixed four-week production plan and the next 8 weeks were laid out for confirmation as the plan rolled forward. I had no finished goods warehousing, product was loaded directly onto trailers and, once full, shipped to the customer. (Contrast that to when I supplied Tesco in the noughties when it was order day one for day three delivery!)

It all came to an abrupt end one day when the customer informed me that they had eight warehouses spread around the city full of my production. The stock was there, he told me, because the changeover times on his major production line was so excessive (it took them twenty four hours to complete a changeover) that it demanded long runs between changeovers to make production economical. He said they about to introduce Just In Time manufacturing and were working with consultants, who were specialists in JIT, to close all eight warehouses. They planned to reduce the changeover time from the current 24 hours to 'minutes' (they said the process was called SMED) and once achieved, they would go close all eight warehouses, reduce the batch sizes between changeover and move to something called Kanbans. The consequences for my company would be much shorter production runs, more flexibility required in production planning and we could only ship product when the Kanbans in their warehouse called for it. Oh, and by the way, no more commitment to four-week fixed, eight week flexible, they would give us a one week forecast of their requirements. Wow! The good news, if you could call it that was that they were prepared to bring us into their supplier programme and educate us in JIT and particularly SMED to help us introduce JIT too. The alternative, they suggested was to 'build yourselves a very large warehouse!'

We never built the warehouse. We introduced the elements of JIT that would help meet their requirements and when they finally brought their changeover time down from 24 hours to 35 minutes (they never achieved single minutes whilst I knew them) we were able to supply to the demand set by the Kanbans.

Three years later, having left the company, I set up my own business called JIT Services, primarily to help other business reduce changeover times. With a targeted letter with the strapline, 'are long changeover times killing your business?' I received a reply from the chief executive of a major UK packaging company based in a nearby town. He asked me to contact with his production director who had submitted a capex to buy a second four-colour offset printing machine. His current machine was working three shifts over five days as well as weekend overtime and, since the new machine would cost £1m, he wanted to be sure it was needed before he signed off the capex.

Needless to say, the production director never got his new machine. Working with the machine operators and using SMED techniques, we dramatically reduced the changeover times and subsequently boosted output such that at the end of a three day exercise, we reduced the manned time of the machine to two shifts operating over five days per week, a 200% increase in output at no additional cost. I well-earned my modest fee!

So where did the term SMED derive from? It is the abbreviation of the term Single Minute Exchange of Dies and was introduced to the world by Shigeo Shingo, a Japanese industrial engineer who worked closely with the Toyota Motor Company beginning in 1969 to introduce a unique method of reducing changeover times. In short:

The SMED system is a theory and a set of techniques to enable equipment setups and changeovers to be carried out in less than 10 minutes.

For the hard pressed production person, there is a seductive attraction to producing in large lots. We talked earlier of economic batch sizes and this was our 'get out of jail card' which gave us the excuse of not worrying too much about the length of the changeover time. After all, it is highly likely that the line crew would have other work to do whilst the line changeover was carried out, usually by the engineers.

There are costs, however, to producing in large lots. These are:

Inventory Costs: Storing what is not sold costs money, ties up company resources and adds no value to the product. It requires people to handle it and space to store it.

Delay: Customers do not necessarily want large amounts of product but often have to order a minimum quantity. Then they have to wait for the full batch to be produced before it can be shipped.

Spoilage: The longer we store things, the more likely the quality is to deteriorate. Ultimately, if it is stored too long it may need reworking or even scrapping. This adds to operating costs and, as important, requires it to be remade again.

So, what are the advantages of reducing changeover times for the company and you as an individual?

Once set-ups can be done quickly, they can be done as often as needed. This means companies can produce smaller lots. This leads to:

More Flexibility: Changing customer demands can be met without excess inventory.

Faster Delivery: Small lots means lower lead times and less customer waiting time.

Higher Productivity: Reduced downtime, since changeovers are always non-value adding, means higher productivity

Higher Quality: Less storage means less storage related defects. SMED allows us to reduce defects by reducing set-up errors and the hit and miss of trialling after set-up.

For company employees, by improving productivity it improves the competitiveness of the company and hopefully job security. Simpler set-ups mean less physical strain and risk of injury. Less clutter in the workplace makes it safer and, with set-up leading to standardised ways of working, there should be fewer tools to keep track of.

Why Quick Change Over?

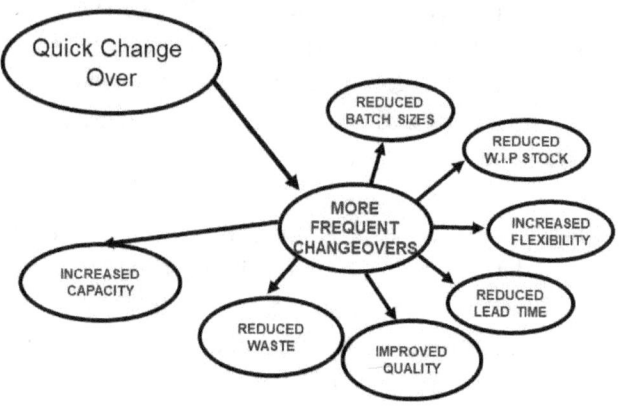

Before we get into the detail of how we introduce SMED, let me share with you a story which inspired me to embrace JIT and SMED. It is a quote I read about a visit by Jack Warner of Omark Industries on a visit to a Toyota supplier...

"In a prep session before the tour, an articulate, knowledgeable Japanese engineer explained to the tour group what they would see. His description of lightning fast model changeovers and zippy, delay free production sounded far-fetched. Eyebrows were raised and interest was piqued.

"The plant tour confirmed that the engineer had not exaggerated. Warner recalls witnessing a changeover on a starter line. About eighty production employees were building small starters for cars. With clockwork precision the crew took fifteen minutes to change over the line to run truck starters four times larger. The line looked to be at full speed immediately.

"...the techniques looked simple to learn and not impossible to transfer to his company."

Source: World Class Manufacturing Casebook, Richard J Schonberger, The Free Press, 1987

Let us start with a definition of changeover:

The time taken to change a process from production of the last good part to the first good part of the next production run.

The time taken includes for example, tool changes and any start up scrap.

This change over time is as, we discussed earlier, non-value added and can be made up of several types of waste.

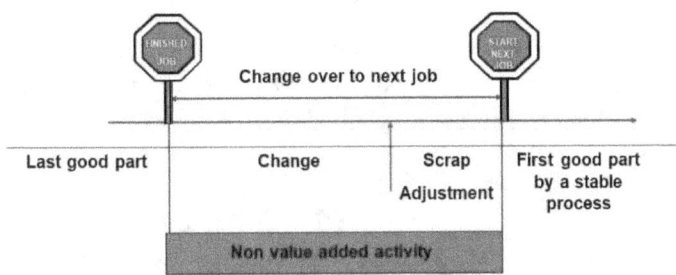

Often, as a consultant when I asked the obvious question, "What is your current changeover time?" if we follow our definition, the answer usually had no relationship to the total time I observed. Many people will exclude setting and running time in their answer which, if the setup has been done badly, can take longer than the time to changeover the machine. We must be honest with ourselves if we are to genuinely achieve SMED.

There are two key elements in any changeover:

Internal Activities

- These must be performed whilst the machine or process is stopped i.e. when not making parts. Examples are die exchange, hose-down of equipment, etc. They are direct contributors to setup time. The more time spent executing these internal activities, the longer the duration of the setup will be. Eliminating, moving (to external), or streamlining internal elements (or creating parallel activities) are keys to setup reduction.

External Activities

- These are the activities than can be performed whilst the machine or process is running and still making parts. They are activities and operations performed while the machine is running (i.e., producing product). These can include tasks such as picking up paperwork, doing a computer transaction, or checking out tools from a tool store. They do not directly contribute to setup time.

Our goal is to record all the elements of the changeover and move as much of it into external activities which can be carried out whilst the process is still making parts.

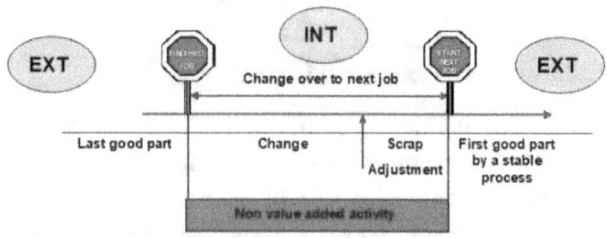

54

There are six steps in reducing changeover time:

1. Select the appropriate set up to reduce
2. Observe and measure the current process
3. Separate internal and external activities
4. Convert internal activities into external activities
5. Reduce internal activities
6. Reduce external activities

Sounds simple doesn't it? In fact it is simple although not necessarily easy. Those that have gone before us, however, have left us with a good roadmap and valuable tools and techniques to use. The process we are trying to achieve is shown below:

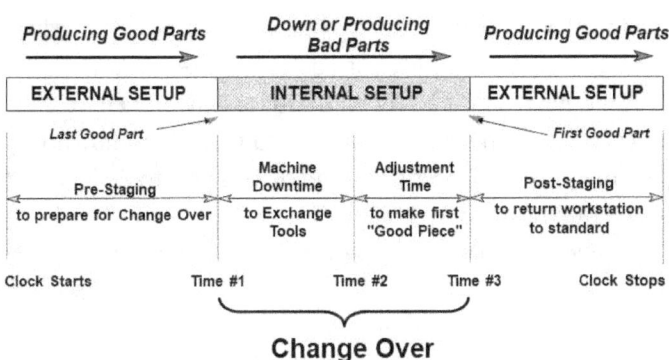

Change Over

Now we are ready to start our changeover reduction project. Before we jump in with both feet however, let us consider how we decide which process to investigate. In his 1984 book 'The Goal,' Eliyahu M Goldratt considers the impact of bottlenecks in manufacturing processes and focusses on how to eliminate them. (A bottleneck is the part of the process that limits the total output of a facility.) He makes the point that an hour saved on a non-bottleneck is an illusion. Whilst this is not totally true, it does make the point that we should focus our efforts on that part of the process that will have the maximum impact on productivity and lead times. When choosing the resource to investigate we should use the 5-Whys we discussed in Book One

Why? What is the need? In other words, what problem are we trying to solve? Is it increased capacity, more flexibility, reduced manning?

When? When do we expect the result to be in place remembering we need to select and train personnel before we can begin? We might for example say that within one year we will have achieved a minimum 50% reduction in changeover time. (I would be amazed if it was only 50%).

What? When I run this as a training session and ask for ideas as to where to start, there is always one machine that people identify as obviously in need of attention. When I ask them to consider the question in another way, what would give us the biggest impact at the lowest cost, it is often a different answer. Until we have a track record of success, we need to establish a policy of low or no cost solutions and therefore choose our target machine or process in the light of this. 40 to 60% of all setup time reductions can be achieved with little or no cost.

Where? We need to identify the bottlenecks in the process. These are the limiting factors in the manufacturing cycle and therefore the places to tackle first. (In the example I gave of my own experience, the print machine I was asked to investigate was genuinely limiting the company's sales).

Who? We need to establish a 'taskforce,' a team of people who are familiar with the machine or process and are able to initiate the changes once they are identified. Operator involvement is the key, although we may want the line engineer involved and possibly other support services.

How? We need to identify potential obstacles and conflicts and overcome them, For example, if there is a union or number of unions in the operation, we need to get them on board. We often move changeovers from a wholly engineering function to one undertaken by the production crew. A communication programme letting the whole organisation know the aims of the taskforce is crucial.

LET'S BEGIN

Whilst set-up time reduction can be done by one individual, it is at its most powerful and most productive when done by a team, preferably from the actual line crew. If it is run on a two or three shift operation, members from both/all shifts should be included. As we discussed in Book One, the members should be familiar with problem solving techniques, process flowcharting etc. to be most effective. The process for the set-up reduction study is as follows:

Agree team roles

It is useful, although not critical to have one member keeping track of the time. Another team member needs to be there to act as scribe. (I found when doing this myself; it is sometimes easier to talk into the camera microphone rather than taking notes.)

Agree and obtain all support materials

Essential to the study is a video camera. I know these days most people have a high quality video camera in their phone, but if changeovers are lengthy, it is better to have a dedicated digital video camera. This can be transferred to editing software for, if necessary, frame by frame analysis (particularly in the adjustment phase.) Needless to say, it is best that the video camera is operated by one of the line operators.

The video operator should focus on the hands of the operator carrying out the changeover. Where the action is away from the line, this should be recorded separately. I remember videoing one changeover with the camera, mounted on a tripod, focussed on the line. The departmental manager asked shortly after the changeover whether he could look at the tape. I set it going and after about 10 minutes into the playback he exclaimed, "Why, there's nothing happening!"

"Precisely," I replied. The operator was actually coming back from the stereo store and was walking as fast as he could! (An obvious part of the changeover that could have been done whilst the machine was running the previous job).

Any tools, fixtures or gauges needed for the changeover should be gathered in advance. Information, such as SOPs (Standard Operating Procedures) or set up sheets should be collected together.

Stage 1 - Review the Current Changeover Method

All setup operations that have not been improved by SMED have four distinct steps, regardless of the type of equipment or operation. The four steps are:

1. **Preparation**, after process adjustments, checking of material and tools for the next run. During this step, materials and parts are removed to the storage area and the machinery is cleaned ready for the next tooling to be fitted. This can be a mix of EXTERNAL and INTRERNAL setup usually accounts for 30% of the setup time. We should aim to make this all EXTERNAL.

2. **Mounting** and removing tools and parts specific to the previous operation and then fitting the parts for the next operation. This, by its description is INTERNAL setup. This accounts for perhaps 5% of the setup time.

3. **Measurements, settings and calibrations.** This step refers to all the measurements and calibrations which need to be made to perform a production operation. It might include centring of a tool, dimensioning and measuring temperatures and pressures. This probably accounts for 15% of the setup time. Whilst this by its nature has to be INTERNAL setup, SMED teaches us to find ways of doing these tasks quickly.

4. **Trial production run and adjustments.** This is the final step of the traditional changeover and is quite often ignored when reporting the changeover time. Making even minor adjustments to the process often depends on the skill of the setup operator. Fine adjustments may be required to smooth out intermittent stoppages of the process or periodic production of defects. This part of the setup process can be as much as 50% of the overall setup time. The goal of SMED is to eliminate this step and make good product as soon as the machine is started.

Once the changeover has been recorded the team need to view it. The first step is to break it down into its component parts, but at this stage in not too much detail. For example, removing a tool might require picking up a spanner and unscrewing four nuts. Rather than focusing on the five elements of this stage, record the total time for the operation. Later, when we brainstorm this we may decide to add two location pins and a quick-release clamp to position the tool. If so, the time to pick up a spanner and unscrew each nut for the tool removal stage is irrelevant.

In this way the team can brainstorm and decide what constitutes the key elements of the changeover. It can be useful, particularly when there is a lot of operator walking from place to place, to record our observations on a Standard Work Combination Table. This can be useful in identifying the value-added part of the changeover (although the overall changeover is non-value adding to the product) and analyse the seven wastes. We can also identify the internal and external aspects of the changeover. Alternatively, we can use a simple changeover sheet as shown below:

Change Over Sheet

No.	Element	Time	Cum.	I / E	╷╷╷╷╷╷╷╷╷╷╷╷╷╷╷╷╷╷╷╷╷╷╷╷

Our initial analysis might look like this:

Analyse all change-over activities

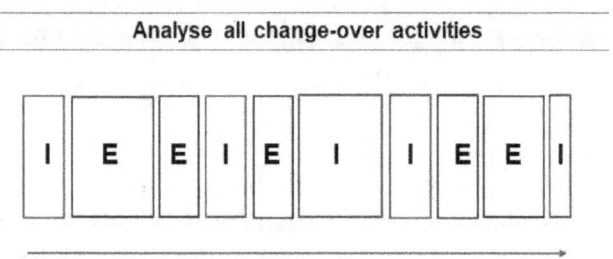

PROCESS STOPPAGE TIME

Stage 2 – Separate Internal and External Activities

Using the current process map, the work combination chart or the changeover sheet, we need to define which the internal and external activities are. One question we must ask when deciding whether an activity is external is whether it is physically possible to be external. For examples some adjustments, which might be considered internal since they are done whilst starting up the new product, could become external by making an intermediate

60

jig that can be set up on a bench and then fitted, already set up for the process.

EXTERNAL OPERATIONS

When considering external activities we must consider which can be done now, for example moving printing plates next to the machine rather than collecting them from the store, or can be moved with minor changes.

We can then define activities that can be moved but only with major changes. This may be highlighted but not implemented, instead to be carried out in a second phase particularly if it requires capital investment.

Remember throughout the exercise that safety and quality cannot be compromised.

When considering external operations, the questions we need to ask to help streamline external elements include:
- What needs to be controlled?
- How can all necessary items for changeover best be organised?
- Where can they best be located?
- How can we control their use (and absence)?
- How will they be maintained in perfect condition?
- How many do we need to be kept in stock?
- Who is in charge of these items?

There is a dark side to simply moving internal elements to either external or parallel elements. That is, the activity still needs to be done and therefore is taking someone's time to do it. This also presupposes that somebody is available to do it. Unfortunately, this is not always the case. Then there is the issue of safety. We all may have heard about the great idea of pre-heating molds before the setup begins to avoid having to wait for them to heat up once they've been installed. Did you ever try to safely handle a 400 degree mould?

Let us now consider methods we can use to convert internal elements to external elements, these include:

1. Advance Preparation

Get familiar with the production schedule (This helps in gathering the correct supplies, ingredients, tools, gauges, etc.). Think about how to prepare, in advance, the equipment, tools, areas, etc. so that this work is performed while producing the previous product (e.g., preheating, re-calibration, cleaning)

2. Standardisation

Minimise the use of different tools, jigs, procedures, supplies to make it easier to work as a team so that others can help with tasks while the machine is running

3. Intermediate Jigs

These are useful in reducing internal operations because alignments, adjustments, and centring can be done in parallel while the machine is running. (Minimises transition to production)

INTERNAL OPERATIONS

Let us now consider how we can streamline internal operations:

1. Use Parallel Operations

Parallel activities are operations performed simultaneously during the setup. This could be two or more people executing different operations of the setup or two or more people working together to execute one operation of setup. As an example, one person clears a line of old packaging material (e.g., cartons from the last product run) while at the same time, another worker puts the new cartons in the feeding hopper

2. Use Functional Clamps

These are specialised hardware that makes use of quick and high leverage movements:

- One-turn methods: U-washers, split-thread bolts, etc.

- One-motion methods: Toggle clamps, cams, springs, magnetic fields, vacuum
- Interlocking methods: Guided cradles

• Eliminate time lost removing and installing bolts

'C' shaped washer	Key shaped mounting holes	Functional clamp

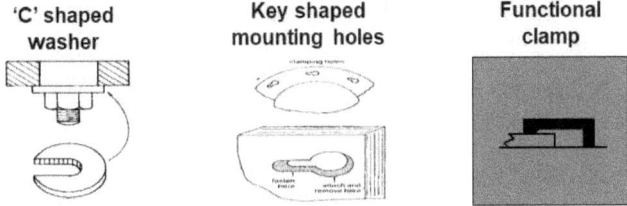

3. Eliminate or Reduce Adjustments:

The following practices are recommended to eliminate or reduce adjustments:

- use of visible centre lines to marry up with a line on the change part
- use setting blocks to avoid measurements
- use physical stops that help position the tool. Often, once the correct position of the change part has been established, a hole can be drilled in the part and a dowel inserted on the bed of the machine
- guides can be added to help position the change part

4. Least Common Multiple System

This involves leaving the mechanism alone and modifying the function of the part. For example, instead of having two different moulds to fabricate plastic parts, use only one mould in which the plastic is directed to one cavity or the other

5. Standardize Operations to Minimize Internal Adjustments

- Instead of operating with different pressures, standardise to one only. This can also apply to sizes and shapes, dies, tools, and jigs. An example of a standard size on a moulding tool is shown below. A shim is added behind each die to maintain the same die height:

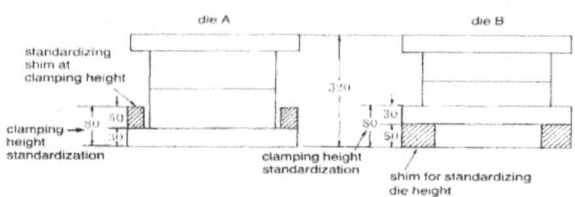

Alternatively, back plates can be added to the dies to standardise the height.

6. Eliminate:

The first thing you should ask about everything we do is "Can we eliminate it?" In many cases in the past, the question we asked was could simply 'do it faster or add ways to reduce the time it takes?' The problem with this approach is that in many cases it leads to better or faster ways of doing things that didn't need to be done in the first place! The real key to this step is eliminating the need for a particular step before eliminating the step itself.

7. Simplify:

If we have exhausted all of our options and absolutely have to do a particular activity, the next step is to make that required step as simple as possible. This is where mistake-proof and fail-safe techniques come in handy. This is also where standardisation becomes our friend because making something simple should also mean making sure it happens the same way every time. An important point to remember here is that the faster we go, the more important it becomes to not make mistakes. Error-proofing is a big part of this step. We consider error-proofing when we discuss Poka-Yoke in Module 5.

8. **Automate:**

   Once we've eliminated unnecessary operations and simplified what is left, we only have one other step to consider - automation. In most cases, we will have dramatically reduced setups by doing the previous two steps. That means that we are probably at a point of diminishing returns by the time we reach this step. Why should we automate then? The best, and some would say only reason, is to gain process control. If you happen to gain something in setup time along the way, that's an added benefit.

In my experience, the vast majority of the benefits from setup reduction can be gained by just using steps – Step 1 advanced preparation and Step 2, standardisation. As you might guess, these two steps usually involve the least amount of cost, as well. How does the process work in the real world? Let's look at some actual examples.

SMED CASE STUDIES

I am sure you appreciate that the subject of set-up time reduction could fill a thousand books describing, in infinite detail, the innovative and varied solutions that have arisen since JIT made its debut in the world. To begin our move to SMED however, we need to keep things simple and look at low or zero cost solutions, if possible. Our aim should be, as a minimum, to halve the change over time. We should also focus on bottleneck resources where the benefit will be evident in better lead times and higher productivity. Indeed, in my business JIT Services, I would tell the client that if I did not halve his or her changeover time, then I would not charge them. It was an easy bet. As I said earlier, companies tend to quote changeover times as the time between stopping the machine and making the first of the new product. The problems of setting and running adjustments are usually ignored. The second reason I was confident was that we are all human and look to the brighter side of life. The setup times I was quoted were usually the best achieved on a sunny day with a tail wind helping! I am going to look at two changeovers, one where external elements reduced the changeover time by over 50%, and one an internal reduction.

External - The Carton Printer

I referred earlier to the first assignment I tackled when starting my own business in 1992. It was a high output, four colour dry offset carton printer used for printing high quality cases. It had already been designed for quick changeovers and had a two man crew who worked very effectively, one at the feeder end and one at the discharge. I show the type of machine below:

Carton Printer

Feed End   Stereo Cylinders

As soon as the machine stopped, the men worked quickly, one removing the old printing plates (called stereos) whilst the other cleaned down the ink rollers. Then one of them disappeared. Not literally of course, he was going to the stereo store to return the old plates and bring back the new ones. (Hence the video shot with nothing happening I mentioned earlier.) Clearly, with the existing setup, he could not leave the machine whilst it was running and to store all the stereos for the shift next to the machine would have been impractical. When he arrived back, he fitted the new stereos whilst the other operator ran up the new ink colours. It was only when they took off the first print samples that they realised that the stereo had been configured wrong, something that would have been spotted by a pre-changeover QC check. They had to spend time modifying it before taking the first off sample. Then one operator disappeared again, this time to have QC check the barcode. (I think you can see where this is going.)

Other changeovers progressed in a similar way, with the operator leaving the machine to collect materials. After the first day it was clear which parts of the operation could be externalised, the question was how to achieve this? I proposed adding a third operator who could QC check materials and pre-kit the materials. A further saving could be made by adding a simple bar code scanner local to the machine. The company had invited the Group Work Study Director to observe my methods, and he questioned where this operator might come from. I explained, over three shifts there were six operators, if we could reduce it to a two shift operation there would be no increase in manpower and lower shift premium costs.

We tried it the next day with impressive results which show below:

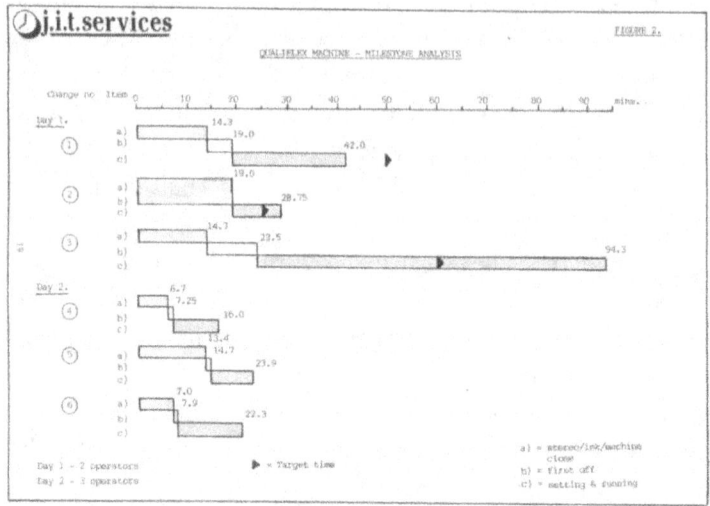

The changeover times were easily halved with very little effort. I advised them to prepare SOPs for the new way of working. One other benefit I shared with them was to stop overproducing. The company had an agreement with the customers such that they could produce up to 10% more than the purchase order. No surprise then that the shift manager took advantage of this and over produced each order by 10%. The planning department, however, did not allow for this and, as a consequence, plans were overrun with the knock-on effect that overtime was often required at the weekend.

So, no fireworks, no magic innovations, just simple application of the principles of SMED. The company moved to two shift working, leads times were reduced, capacity increased but, sadly, the production director never saw his new machine. I, on the other hand, was invited to do further studies to help improve productivity elsewhere.

Internal - The Injection Moulding Machine

The second study concerns a tool change on a 500 tonne injection moulding machine. If you are not familiar with injection moulding, let me describe briefly the process. The product is created in a mould which has two halves; the gap between them forms the shape of the product. Hot plastic is 'squirted' into the mould which is then cooled with chilled water before the press opens, and the product is discharged. The machine in question was large by comparison to the majority of moulding machines and was used to produce large injection moulding containers. On a good day, a tool change would take one man four hours to complete. Since it was required to produce 168 hours a week, reducing changeover would have a major benefit.

The initial layout of the tool is shown below:

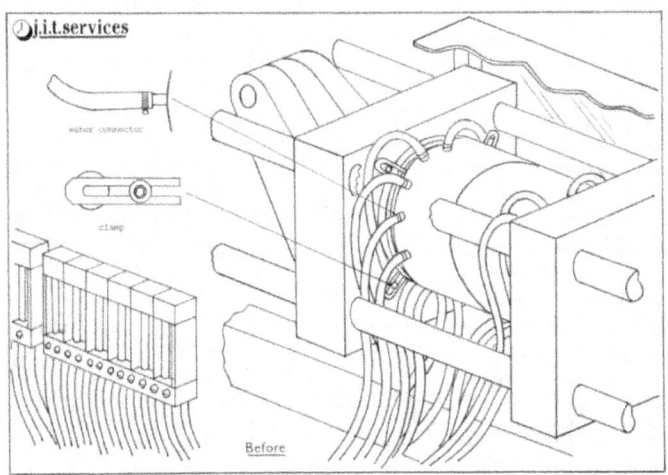

The tool, weighing over a tonne, was clamped onto two 'plattens' using 'U' shaped clamps. It was located by trial and error and once positioned the operator connected and tightened the water pipes using 'Jubilee' clips. All very time consuming as you might expect and ripe for improvement.

The final layout is shown below, the time taken being reduced to one hour.

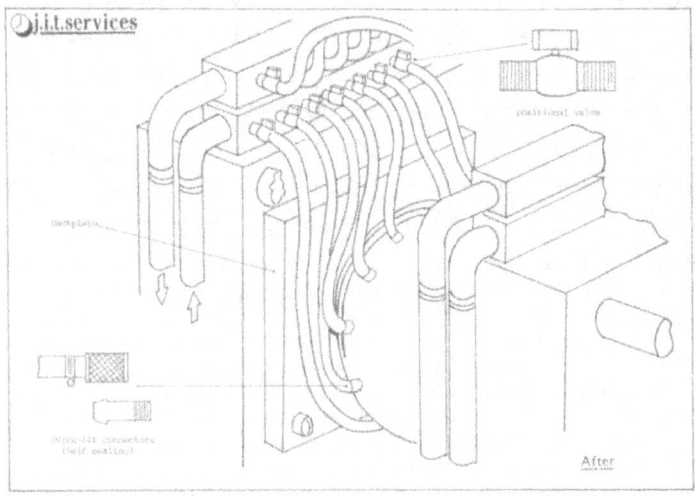

Instead of the jumble of water pipes coming from the flow manifold, two manifolds have been added to the top of each platten. The pipes have ball valves which close in a quarter turn. Quick-fit connectors have been added to the other end of the pipe with a male connection on the mould itself. The connectors are self-sealing. The two halves of the mould have been mounted on back plates which require four bolts to secure and, by their nature, automatically position the mould. All are simple, low cost actions that have a considerable effect on the overall changeover time.

SMED – IN SUMMARY

We have covered many of the areas of setup time reduction that, when combined, can give a dramatic reduction in changeover time. I summarise them below:

70

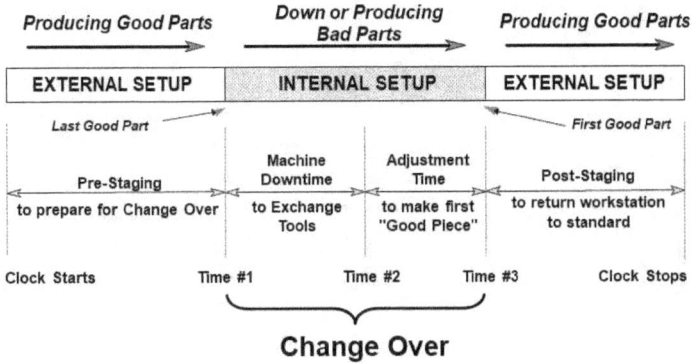

Change Over

1. Identify critical machine or process
2. Select the team
3. Training team in the techniques of SMED
4. Establish team goal
5. Observe, map and record current practice
6. Identify Internal and External activities
7. Eliminate adjustments
8. Brainstorm solutions
9. Apply and test solutions
10. Streamline remaining actions
11. Record and monitor new performance levels
12. Train operators in the new standards
13. Look for standardising in the plant

MODULE 4

INTRODUCING PULL

PRODUCTION CONTROL OPTIONS

This is the moment in our book when we seriously have to consider giving up our comfort blankets! You may already think that creating flow has given us challenges. (Evidently we no longer seem to have problems; the psychiatrists say the word is too negative) Placing independent machines in line and exposing faults means that these can no longer be ignored. So before we introduce the subject of PULL, let us consider the way many, if not most, manufacturing operations schedule production. It's known as a PUSH system.

Push System

A push system produces parts according to a forecast of customer demand. The forecasted demand is usually either greater than or equal to actual demand. In a Push system, each process in the manufacturing chain operates as an "isolated island", producing parts according to a forecast-based schedule received from production control, without regard to the actual demand of the downstream process (customer). Often companies have gone to great lengths to introduce computer based systems to manage this complexity; MRPII and SAP are widespread. The push system operates under the common misunderstanding that all machines must run at maximum capacity for the entire production schedule. As a result, inventory levels in a Push system are difficult to manage and usually greater or less than optimal.

Planned production

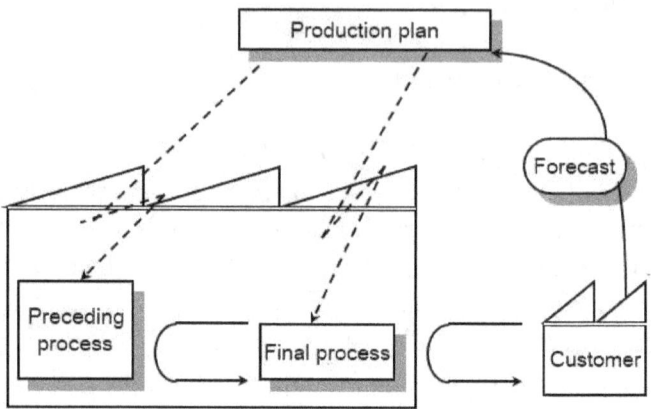

Think of a Push system in terms of pipework and the steps in manufacturing as the water flowing through it:

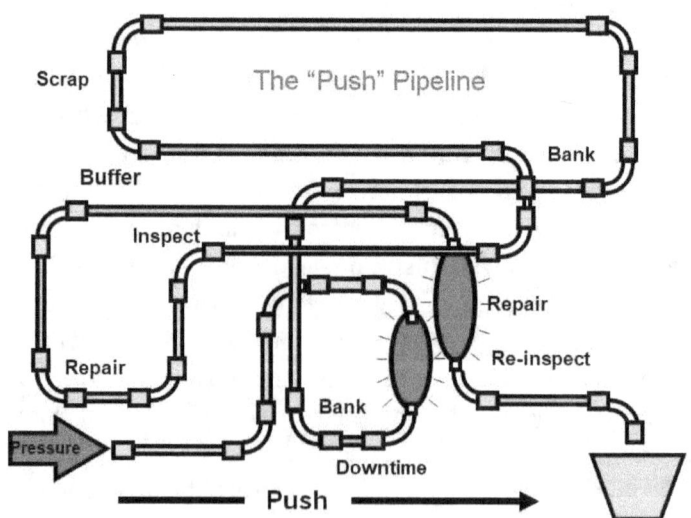

In this analogy:

- Pressure builds at the end of the process

- We all become expeditors, especially when we get blockages and WIP starts to build
- We operate in a reactive mode
- Emphasis is on fire fighting
- Causes of stoppages and defects are covered up
- Focus is on product vs. material flow
- We build safety stock, equipment, facilities, etc.,"just in case" we need them
- Emphasis is on who is 'screwing up' rather than why it is happening
- Waste is a way of life, but hey-ho, there's always rework!

PUSH production depends on *production work orders* which identify the type and quantity of work to be done. It is often generated by a planning system called MRPII which evolved from the original system, MRP. MRP stands for Material Requirement Planning and this became Material Resource Planning (MRPII) to also cover the labour requirements.

MRP depends on a Master Production schedule which is rigorously maintained. Planned and actual output is compared daily and discrepancies addressed weekly through changes in the master schedule.

Pull System

A pull system produces parts only in response to actual customer demand. In a Pull system, parts are produced only in the amount of what has been consumed by the downstream process (customer). Inventory levels in a Pull system are easily managed, and are calculated to enable demand to be met without excess inventory.

In our pipework analogy, it looks something like this:

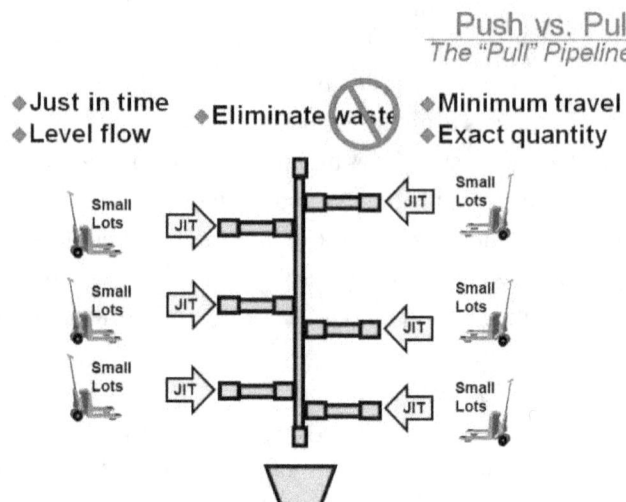

The PULL in this case is created by gravity creating a Venturi effect to pull the 'water' through the pipework to the container (customer) beneath.

Before we explore the details of PULL production, it is worth noting that if a company has one product with no variations, or alternatively has seasonal or steep fluctuations in demand, pull may not be of benefit. It benefits more those companies with a variety of products with demand relatively constant throughout the year. All the other elements of lean however, including waste elimination are still of value.

Now is the time to introduce the term Kanbans. Just as a *production works order* serves as the initiator of work in a Push system, Kanbans serve as the *production order* for the pull system. We will discuss Kanbans in detail as we progress. The key difference to highlight between push systems and pull systems is this:

Order information in a pull system – the Kanban – travels upstream from sales to assembly to procurement instead of,

76

as in a push system downstream from planning to procurement to assembly to sales.

Wow! At first sight it seems quite daunting. So the warehouse or, if there is no warehouse, final operations get the order and then what? They haven't any stock to despatch. Usually they will have seen the stock appear and then a despatch note is raised to initiate delivery to the customer. In practice, however, the order is still received by the sales department and it is the flow of information at first that changes. We will expand on this later, but the diagram below illustrates the difference between the two systems:

Customer order production

Produce parts needed in the exact amount needed, based on customers' firm orders.

Manufacturing with minimal inventory, without warehouses and stores, Just-In-Time Production.

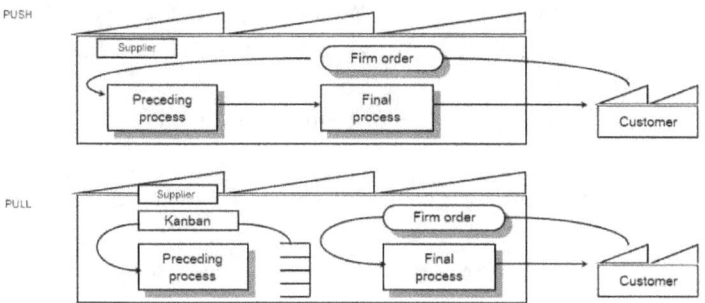

It is worth noting here that for true once-piece flow Kanbans are not required other than to initiate production from the customer. Any WIP added to mitigate against interruptions is considered separately.

Before we move on to discuss Kanbans, let us compare the two methods of production control and consider their similarities and differences.

Similarities

Both the reordering point method and the Kanban system:

77

- Enable inventory to be managed without paying attention to demand fluctuations.
- Both systems are not suitable when there are frequent sharp demand fluctuations
- Both help keep inventory costs down
- Are conducive to use in an automated recording system

Differences

1. *Information and Goods*

   In the reordering system, information and goods are kept separate from each other and inventory is managed by a separate warehouse system

   In Kanban systems the goods and information are kept together

2. *Management*

   Reordering point systems require constant inventory management through procedures such as issue notes to move goods to line side

   Kanban on the other hand requires no management

3. *Visual Control*

   Reordering point systems do not lend themselves to Visual Control as we introduced in Book One. It relies on the information in the MRP system being up to date and accessible through a computer terminal.

   Kanban enables Visual Control

4. *Factory Relationship*

   Reordering point systems are managed separately from the factory and information is usually transferred through PC terminals or computer print outs

   Kanban, by its nature, is closely linked to the factory and factory operations

5. *CI Activities*

Reordering point systems tend to be static with little drive to improve the situation

Kanbans on the other hand, are obvious in their use and drive Continuous Improvement efforts to reduce their number

KANBAN

A Kanban is a pull signal that is used to alert an upstream process that it needs to produce a product or a batch of product that has been consumed by the downstream process or has been withdrawn from the supermarket that exists between the two operations or with an external supplier.

The Kanban system determines the production quantities in every process and because of this it has been called *the nervous system* of lean production. Its primary purpose is to stop us overproducing by producing only what is ordered, when it is ordered and in the quantity it is ordered. It is essentially a *works order*. The difference to the 'push' system of manufacturing is that the *works order* moves with the material.

You are already familiar with Kanbans and Kanban systems, although you may not know it. Whenever you go to the supermarket and 'pull' an item from the shelf, you are using Kanbans, or more accurately, interfacing with the supermarkets Kanban.

Have you noticed how for most items that are on shelf are packed in a carton (known by the supplier as 'shelf-ready' packaging), containing maybe 6 units? Once a carton is empty it leaves a space on shelf. This creates a pull single for the shelf stacker to replenish it with the same amount, which is, one carton, six units. The more units a supermarket sells of a particular product, the more space on the shelf they will allocate. (Supermarkets don't sell products, they sell shelf space and the return per square meter will ultimately dictate the range and quantity of products they offer for sale.) If a product sells well, they may allocate say twenty spaces on shelf, if a slow seller, maybe four. As the shelf stacker depletes the stock in the stores warehouse, an empty pallet will then create its own pull single for the supplier to replenish that pallet. Although the sale in also tracked electronically at the till and this will automatically reduce the system stock for that item, in actual fact the Kanban system is a more accurate determiner of the amount of stock to replenish, since till sales do not see breakages, spoilt stock or thefts.

The other Kanban you may have used is if you were purchasing a high value item. The one on display is likely to be secured so it cannot be removed and, instead of a box containing the product being on a shelf below the item, as in low value products, there may be a card holder next to it. In this will be a number of cards which will have printed on them the item description and a note saying, 'Please take this to the till' (or perhaps Customer Service). The person at the till will take your card and bring you the product. The number of cards in the system relate to the amount of stock being held. The card they have asked you to bring to the till is a Kanban card and usually has the details of the product you are buying on it, the information being for the store than yourself. This card will not be returned to the shelf. As more stock is sold, more cards are taken to the till until the number of cards brought to the till triggers a reorder of the product. Only when the stock is available will the total number of cards on the shelf be replenished. We will see more detailed examples of these types of Kanban as we progress. Interestingly, the term 'supermarket' has been added to the lexicon of pull and is a term used for controlled inventory between two operations that belongs to the supplying process. It is used where continuous flow cannot be achieved.

The two major types of Kanban are Production Instruction Kanban and Withdrawal Kanban. We will also here other terms such as Signal Kanban and Parts Kanban but these are variations on a theme and their purpose is either to initiate production or to withdraw parts from a store or 'supermarket' or from a supplier. A simple schematic is shown below:

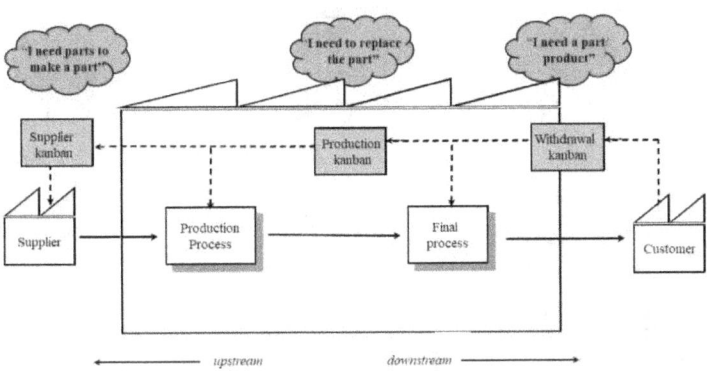

Sometimes a Kanban loop only requires the use of a Production instruction. One example where this might happen is if the customer is the shipping department and shipping is using a pick list to pull parts from the Supermarket. Alternatively the customer is simply pulling product from us on the basis of need. The important part is that the supplier must receive the replenishment Kanban to produce in a timely fashion from the Supermarket. The production instruction Kanban will always stay on the product until it is withdrawn from the Supermarket. When the product is withdrawn the Production instruction

Kanban is sent to the supplying operation. Let us add detail to our simple schematic to illustrate this.

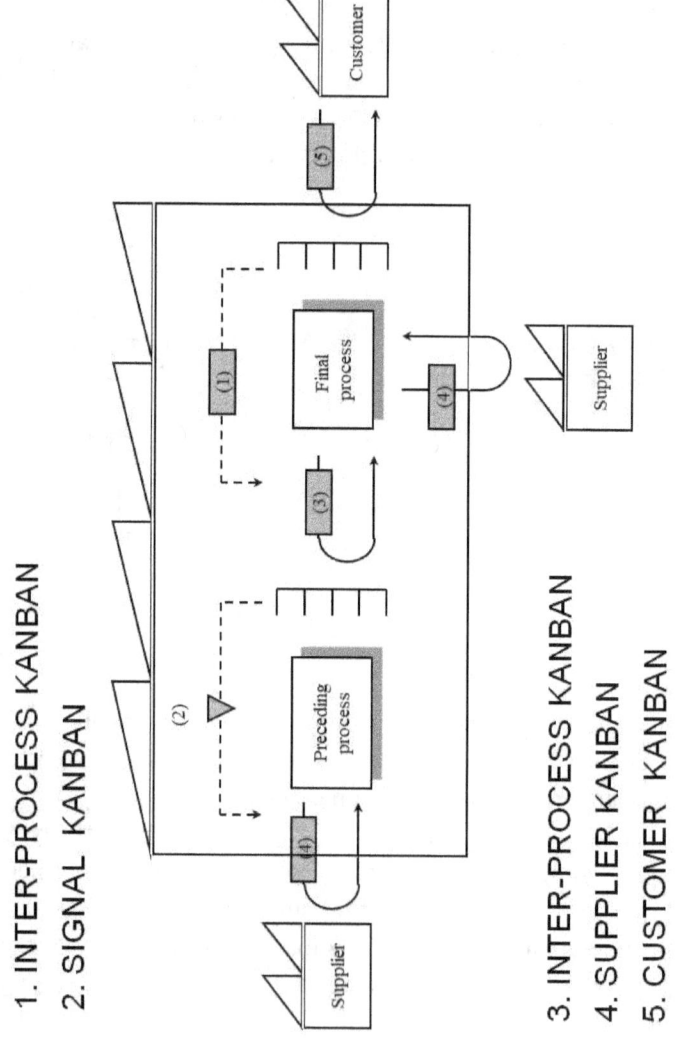

1. INTER-PROCESS KANBAN
2. SIGNAL KANBAN
3. INTER-PROCESS KANBAN
4. SUPPLIER KANBAN
5. CUSTOMER KANBAN

So, let us look at an example of a Production Kanban shown in the diagram, and the information it contains

DATE	5EY40LAZAB
Steering Column Cover	

CUSTOMER:	PACK SIZE: 24
TRIGGER POINT: 25 CONTAINERS	PACKAGING RETURNABLE

STORAGE LOCATION: WJ FINISHED GOODS	CARD NUMBER 1

It is giving us information on the part required, the type of packaging and the trigger point, all the instructions needed for the production process. As we said earlier, it is our *works order*. As the part is removed from the supermarket, the card is removed and sent 'upstream' to the manufacturing process and is their instruction to produce a lot as described on the card.

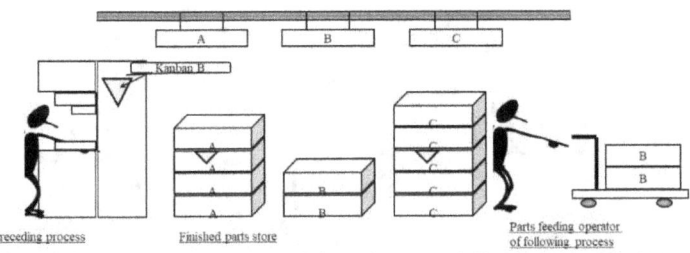

The Kanban stays at the machine until the lot is complete and then it is removed and added to the pallet. It moves with the pallet to the store or supermarket and remains there until it receives a pull signal from the downstream process. And so, as it

is removed from the supermarket, it is sent upstream, once again, to initiate the production cycle again.

A one-card system should be used when the upstream and downstream processes are close to each other. Separate storage of inventory at the downstream process is not required. In this case, the removal of product from the supermarket by the downstream process results in a production kanban being sent to the upstream, or supplying, process, at the same time.

When upstream and downstream processes are located far apart from each other, an additional inventory store is required at the downstream process. In this case, a two-card system is used. The second card is a withdrawal kanban, which is used by the downstream process to control the level of line side inventory. The withdrawal kanban is attached to each container at the downstream process. As product is consumed from the downstream store, the material handler takes the withdrawal cards to the upstream store and retrieves an equal amount of product, removing the production Kanbans and attaching the withdrawal Kanbans to the product as it is moved downstream.

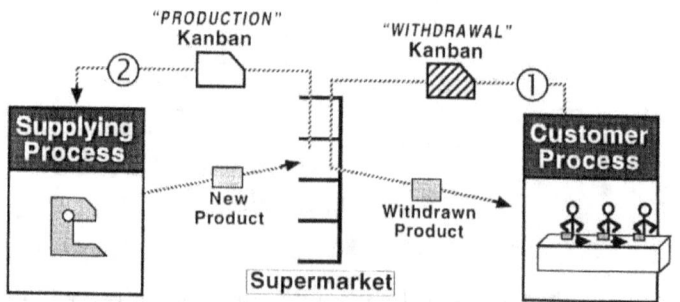

So, to summarise our discussions so far, we have basically three types of Kanbans, Production Kanbans, Withdrawal Kanbans and Supplier Kanbans:

Every other type of named Kanban is a variation on these themes. In the following process we have what is called a *Pickup Kanban*, the purpose of which is shown below:

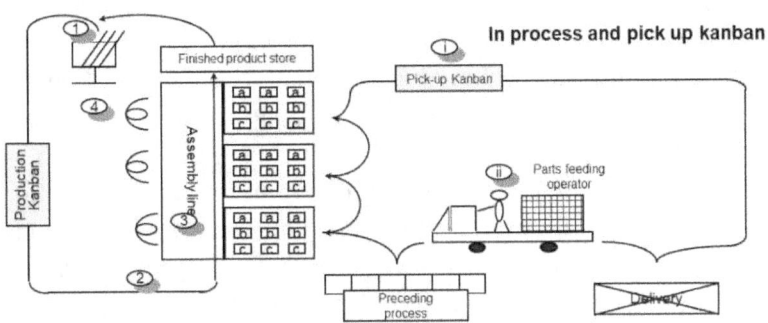

The sequence of events is as follows:

1. A worker hands a *Production Instruction Kanban*, which has been removed in the finished product store when the product was shipped, to a worker at the head of line. This is the PULL signal for production.

85

2. The worker at the line takes necessary parts from the supermarket in the order of the Kanban and assembles them. The worker then sends the work piece with the Kanban onto the assembly lines.
3. The assembly line workers assemble parts according to the instructions on the Kanban.
4. The worker at the back of the line puts the Kanban on the finished product and places it in the finished product store.

Our diagram shows the supermarket with a mix of parts a, b and c used in the assembly line. Each part has a *Pickup Kanban* attached.

As the parts are used in assembly line, the parts feeding operator collects the *Pickup Kanban* when it is removed. This is the PULL signal to withdraw components from the preceding process. This is shown at stage (i). The parts feeding operator collects the necessary parts from the preceding processes or the purchased parts stores and fills up the assembly line supermarket according to information on the received Kanban (ii).

In this way, production is initiated as finished product is shipped to the customer. Information flow is contained on the Kanban cards.

Contrast this to the PUSH system where information is contained on the production plan and the works or production order. *Production orders* are raised by the scheduling department every 24 hours. The production plan, derived from the Master Production Schedule, MPS, drives the sequence and quantity of production. Components are issued to lineside by the warehouse according to the bill of materials for the current order. Finished products are delivered to the finished goods warehouse at the end of the production run whether they are to be despatched or not. Unused components are returned to the warehouse using an issue and return system.

Kanbans can work with MRP systems. Since a Kanban is a pull system, a number of *Production Kanbans* can be issued, as the production *order* to the process closest to despatch. The Kanban cards can be held in a container which is filled up at the start of the shift. If we are implementing Kanbans for the first time, this is a good way of starting. The aim, once the flaws in the system have been ironed out, is to reduce the number of cards in play.

This moment, perhaps, to introduce a special kind of Kanban called a *Signal Kanban*. As its name suggests it sends a signal to the process that something needs to change. It is quite distinct from other Kanbans in its shape. It is triangular. It can be used to signal a changeover is required or can be used for re-ordering. A sample card is shown below:

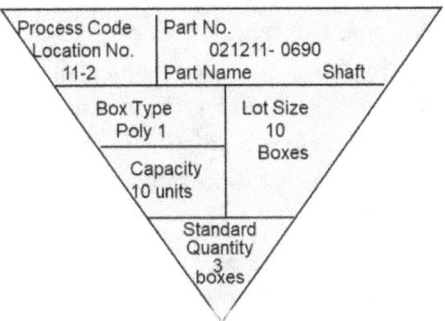

It contains all the relevant information required for production. When used for stock control, it is added to the pallet at a level just above the safety stock level. As stock is consumed, the pallet level lowers until it reaches the re-order point. This is identified by the signal Kanban.

In the example, the reorder point is when stock has been consumed to the last layer of boxes. The Kanban is removed by the warehouse operator and taken to the supply process to initiate the production needed to replenish the stock. Signal Kanban can also be used on a type of planning board called a Heijunka board to give a distinctly visual signal that a changeover is due. We will discuss this in more detail when we consider levelled flow in Book 3.

Kanbans, however, do not always have to be cards. Sometimes a simple square is enough. One example of this is FIFO Kanbans. We talked earlier of creating flow by placing each step of the manufacturing process in line. All very well until of course, one of the machines breaks down. Rather than interrupt the whole flow of the line we can add space for WIP before each machine. But how much? We can calculate this based on the average lost time we have experienced in the past. Since we are still operating a PULL system we cannot afford unlimited WIP before us as we will never have time to process the backlog. This is where the Kanban square comes in. We don't need to provide the process with information, it is already known on the production Kanban at the end of the line. We just need a square which if empty, can be filled by the supplying machine. Hopefully the breakdown will have been resolved before we fill up all the Kanban squares. If so, the machine processes the waiting stock through FIFO – First In First out. If the machine is still down, the supplying machine has to stop; there is no longer a pull signal.

HOW MANY KANBANS?

There is a formula for calculating the number of Kanban that are needed. It is a function of the daily output and lead times together with the pallet quantity. Our deliberations, so far, have been for one purpose. To produce in smaller lot sizes. With quicker changeovers, inline flow and pull, we can make product in smaller batches and more frequently. This should drive down our pallet sizes as well as our overall level of stock. The relationship between these factors is as follows:

No. of Kanbans = <u>Daily Output x (lead time + safety margin)</u>
<u>Pallet Quantity</u>

Where:

- Daily output = <u>Monthly output</u>
  Workdays in month
- Lead Time = Manufacturing lead time (processing time + retention time) + lead time for Kanban retrieval
- Safety margin: Zero days or as few days as possible
- Pallet capacity: We should try to keep pallet contents small and instead increase the number of deliveries.

It is probably becoming very clear to you, my reader that, whilst the use of Kanbans is an obvious development of one-piece production and pull, the variations in use across the many manufacturing operations can be almost infinite and at the same time very specific. I am assuming however that, since my book series is called Lean Made Simple, my reader is looking for an introduction to the elements of lean, explained in an easy to understand way, rather than chapter and verse of how to develop Kanban systems in his or her operation. There are other types of Kanbans identified, Emergency Kanbans and Express Kanbans for example. There are other terms you will come across such as milk runs and water beetles. I will leave it to others to explain these.

If you genuinely want to know more, I cannot recommend too highly a series of books created by the Productivity Press Development Team designed for use on the shop floor. The book I would particularly draw your attention to is 'Kanban for the Shopfloor' published by Productivity Press. Of more use now, I believe, is to offer you an example from my own experience of how Kanbans have been introduced to create pull.

KANBAN CASE STUDY

So you want to introduce lean but where do you start? Clearly, we know from Book One that you need to have created stability in your business. I was an associate consultant with a Lean Consultancy that had been asked to help an international chip manufacturer introduce lean into their operation. As you may imagine, the manufacture of silicon wafers is a complicated, high-tech process with some wafers requiring over 150 separate operations before they become the silicon chips we see in microprocessors and computers. Controlling the flow of wafers through the plant, known as a FAB, (wafer FABrication) was a sophisticated computer system which the operators interfaced with at each stage of the manufacturing operation. Moving to a card based system at the start was clearly 'a bridge too far'.

As we now know, we pull from the operation closest to the customer; the obvious place to start was at the end of the process, known as Final Ops. 'Final ops' was the finishing stage of the manufacturing process and also included a test of each individual wafer. Our first task was to calculate the Takt time for each product so that we could set the drumbeat of the operation. The variety of different wafers that was produced was mind-blowing but the principals are the same when calculating Takt time. The only difference to what we have learned so far is that we needed to operate with mixed-model Kanbans in the FAB. This is something which is outside of the scope of this book. We will meet this again in Book Three. Once we calculated the Takt times, based on the pattern of actual customer demand, we knew the drumbeat for the entire FAB. We could then use the computer system to 'pick' which wafer to pull.

So, how did we determine the rate the operators needed to pull wafers through the FAB? Clearly, there were no Kanban cards to use as a visual signal to manufacture. Still wishing to keep a visual system for setting the rate of Pull, we introduced the operators to PACE charts. A Pace chart, as the name suggests, sets and displays the pace at which we need to pull wafers to the resource. We chose the Back Metals tools as the ones to set the pull rate of the wafers. A sample PACE chart is shown below:

PACE Chart

Date _____

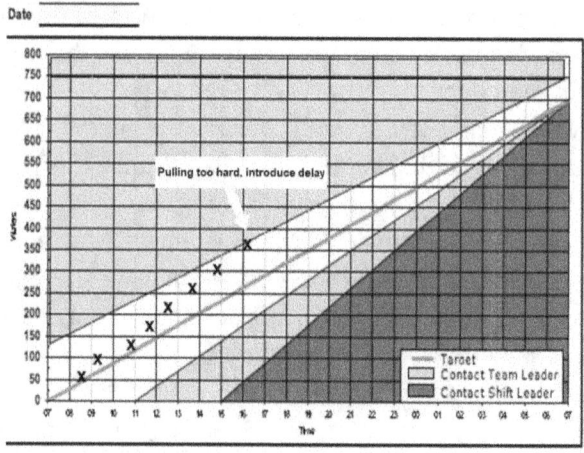

To explain, the operator processes a wafer and adds it to the chart. The figure added is the cumulative production for the tool for the 24 hour period which is manned over three shifts. The Target to achieve is shown as a green line and shows roughly when the next wafer needs to be processed. As the shift progresses it shows clearly the progress so far and the trend, whether we are keeping to the target line or not. If the operator is pulling too hard, then it triggers an action; in this case, to introduce a delay. (This will eventually stop the downstream process as the Kanbans before the tool will not clear.) If the processing rate falls below target it triggers an action to escalate the problem to the team leader and ultimately the shift leader.

You may have expected that we would place the control of the pull at the very end of the process. We considered this but the final process of testing did not lend itself particularly well. Some wafers were tested within minutes, others took hours. Instead, we placed the PACE chart at the resource before and used FIFO Kanbans to pull work through testing. In practical terms, when a wafer finished on the test machine, it was sent to packing and then dispatch. As the test machine became available, signalled by an Andon light, the operator pulled the wafer that had arrived first at the WIP store (supermarket). Here is the flow of work and information:

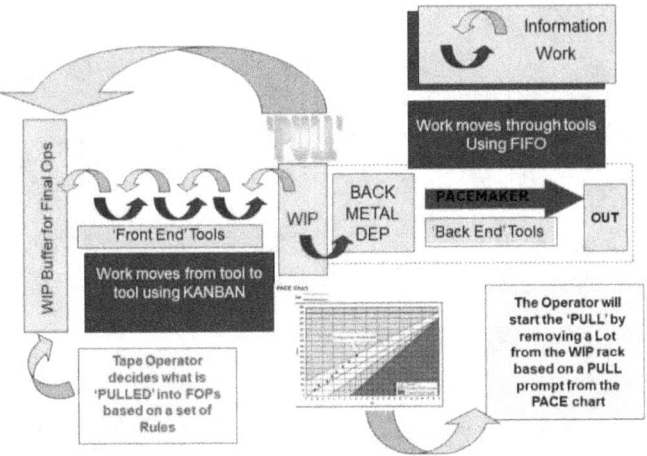

Once we had set the pace for pull, the next task was to set up the Kanbans. I have given you a method for calculating the number of Kanbans but, in this case, the process was too complicated. Instead we set up a simulation by mimicking the tools in the training room. We arrived at the number by starting high and gradually reducing them when it had an adverse effect on flow. Since we knew the cycle time for every tool, we could mimic what the plant would do. I recommend this as a good way to test your theories if you are planning to introduce flow and Kanbans. It is risk free and, if not fool proof, a good way to start. (If your processing times are long, scale them down by a factor of ten i.e. 60 minutes becomes 6.)

A schematic of the process is shown below:

Kanbans - Schematic

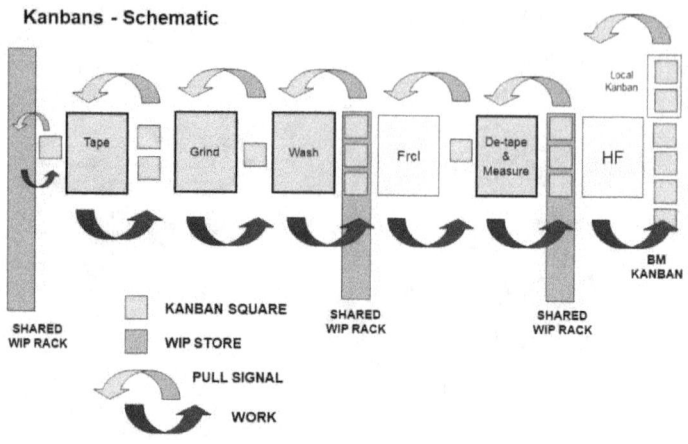

Once we had agreed on the number of Kanbans in the simulation, we set up the plant in the same way and began to train the operators on how it would work. We produced simple diagrams to illustrate 'how' to pull. The 'what' to pull came from the computer system, as previously, the operator selecting the lot that has the earliest despatch time.

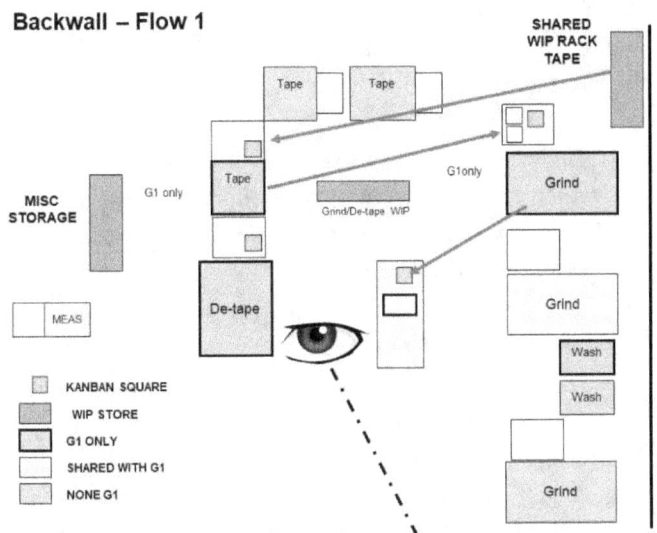

Backwall – Flow 1

SHARED
WIP RACK
TAPE

Tape   Tape

Tape   G1only   Grind

MISC
STORAGE   G1 only   Tape

Gnnd/De-tape WIP

MEAS   De-tape   Grind

Wash

KANBAN SQUARE

WIP STORE   Wash

G1 ONLY

SHARED WITH G1

NONE G1   Grind

Once the principals behind the new method of working were explained, the operators readily took to the new system. The managers, however, needed more persuading. They tended to see their jobs as expediters, (to be fair, there was so much congestion in the rest of the FAB that this was always needed) so when a tool went down with a problem and the flow temporarily stopped, they tried to override the Kanbans. Not happy with seeing the upstream tools doing nothing, they began to build WIP near to the tools. A few harsh words by the FAB General Manager soon cured them of their indiscretions!

As you can see from this example, Kanbans do not have to be all singing, all dancing. Their main purpose is the initiate the pull sequence. A simple square on the floor is sometimes all that is needed, even in non-lean operations, especially for example is there is only one product. When the square empties, it acts as a signal for the service operator to bring another pallet.

KANBAN – RULES

Having now seen Kanbans in action, let us remind ourselves of the rules governing their use:

- *The consuming process consumes only what's needed*
- *The supply process produces only what the kanban pull system authorises, when it authorises it*
- *Defect products will not be sent to the consuming process from the supply process*
- *Kanban should reflect changes in demand as demand is frequently subject to trends*
- *The number of Kanbans should be minimised over time*
- *The rules are none negotiable*

I said in the introduction to this book that a disciplined approach is required to operate lean, particular when introducing pull and Kanbans. If we stick to these rules and not try to override them like the supervisors in the example I gave, then it will pay dividends. The key is to think big but start small, prove pull and Kanban can work in one part of the operation and it will spread, as other areas realise the advantages.

SMART MOTORWAYS

As a postscript to our module on Pull, let me explain my reasoning when, in the introduction, I said that 'smart' motorways use lean principles. A normal motorway is very much a 'push' system; drivers join the motorway free and unhindered to set their own pace of travel, hopefully within the legal speed limit. Normally, when the motorway is relatively quiet drivers have no problem in choosing their speed or lane. Think of the three lanes as three separate lines of inventory and the act of driving along a lane as a production process, consuming this inventory. When the traffic starts to build up, inventory reduces and cars start to change lanes, looking to consume the inventory of the new lane. This happens particularly when they hit a bottleneck in their current lane, often caused by a slower car ahead or slow-moving trucks traveling up a hill. Since there are no restraints to hinder them, drivers are free to choose their speed and lane, sometimes weaving in and out in the hope of keeping moving at their preferred speed. (Many drivers are like production people, they want to process as much inventory as possible!) Eventually, as traffic builds and lanes start to become filled, the motorway has less and less 'inventory'. Finally, with cars bumper to bumper, the motorway is to all intents and purposes, blocked. Only the few vehicles at the front can move.

Clearly, when traffic is excessive, we need to 'balance' the usage of the three lanes as we balanced our production line in Module 1. On a Smart motorway we achieve this by setting the 'Takt time' we want the cars to travel at. (The traffic manager operating the overhead sign is effectively the 'customer' of the motorway system and sets the rate of pull.) This is the purpose of the overhead speed-limit sign with a built in camera; to set and monitor speed. By setting a fixed speed, it balances the flow in all three lanes to the same level. The more traffic, the lower the speed. There is now no longer a need to change lanes; the traffic in the other two lanes is moving at the same speed as we are. At lower speeds, vehicles can move closer together, reducing inventory if you like, and allowing more vehicles to use the lane. Finally, instead of more cars pushing their way onto the motorway, we 'pull' them by using traffic lights on the entry slip roads. We pull at a rate that does not cause the congestion we would normally see at junctions – no longer do we need to change lanes to accommodate other vehicles joining.

Don't get me wrong, this isn't a true lean system but the lean principles of flow, balancing and pull are there to be seen.

MODULE 5

MORE LEAN TOOLS

In Book One we learnt of the 7 Tools of Quality and other problem solving techniques including the 5-Whys and Fishbone Charts. These are essential steps in our journey to create stability. Now that we have created flow and one-piece production, process interruptions and waste are more critical than ever. Before we take the next major step and introduce Value Stream Mapping, it is useful to look at a number of techniques to help us minimise disruptions in our operations and continue our progress towards zero defects.

JIDOKA

We have mentioned the Toyota Motor Company in our deliberations without going into too much detail so far. Toyota, and in particular a man called Taiichi Ohno, are responsible for developing lean as we know. When they introduced it, it was known as, and still is, the Toyota Production System (TPS). It was Ohno who gave the world JIT and the subject we are going to consider next; Jidoka.

Traditional car manufactures in the 1960s had a phobia about production lines; they had never to stop. Time, we are told is money and a minute lost on the line was a car lost. So, any problems with quality were brushed aside to keep the assembly line running. The philosophy was, 'produce now, fix later'. Rather than follow the Western logic of producing lots and lots of similar types of vehicles, Ohno wanted Toyota to be able to produce small numbers of many different kind of automobiles. It was he who devised the 7 wastes and he fixated about reducing these. The worst waste, in his view was overproduction and to tackle it he developed Just in Time manufacturing and Jidoka. So what is Jidoka?

Jidoka is a Japanese term that translates to something like 'automation with a human mind'. Or, put another way, the autonomous control of defects. It is the ability of a machine or production lines to be stopped immediately in the event of an abnormality thereby preventing the production of defective items. Clearly with one-piece flow, the emphasis has to be on avoiding defects progressing to the next stage of the process.

Automation, of course has been around for decades, if not centuries. Machines, producing automatically, traditionally had a machine minder to check things were operating as planned and stop and sort any issues as they arose. But we now know, as Ohno knew, that watching machines does not add value. What he wanted was an intelligent machine that could tell the operator when it had a problem.

The concept of autonomous machines is not new. The concept of Jidoka originated in the early 1900s when Sakichi Toyoda, founder of the Toyota Group, invented a textile loom that stopped automatically when any thread broke. Previously, if a thread broke the loom would churn out mounds of defective fabric, so each machine needed to be watched by an operator. Toyoda's innovation lets one operator control many machines.

In Japanese, Jidoka is a Toyota-created word pronounced exactly the same (and written in kanji almost the same) as the Japanese word for automation, but with the added connotations of humanistic and creating value. One of the key benefits is to involve humans in the decision-making process.

There are three tools we use in Jidoka, one of which you will be pleased to know that we have already covered; Andon: The others are:

- Full work system
- Poka Yoke

Full work system

What do we mean by a full work system? It occurs when a supplier process stops feeding the downstream process when it is full (for example when the desired numbers of parts have been provided). We have covered this, of course, when we discussed pull and Kanban. It also involves the creation of full standard work content for operators – eliminating the possibility for variation and mistakes.

So, let us consider Poka –Yoke.

POKA- YOKE

This is a method of mistake proofing to avoid unnecessary waste and move us further along our path to zero defects. Let us define it as:

Methods that help the operator avoid mistakes in their work caused by choosing the wrong part, missing a part, installing a part incorrectly and other human errors

For all of its strange sounding name, Poka-Yoke is something we are all very familiar with and it plays an essential part in our day to day lives. For example, the petrol cap on my wife's car is opened with the ignition key. (Yes, sorry, I know modern cars now have buttons!) If you try and remove the key from the cap, it doesn't work. You can only remove the key once the cap is locked again. Before it was introduced, believe me, I have driven away with the petrol cap still balanced on the petrol pump too many times to tell! When we leave the microwave door open have you noticed that it won't start? These are two examples of Poka-Yoke in action.

Shigeo Shingo, our man from SMED, first introduced poka-yoke to Toyota in 1961. It was originally called baka-yoke (fool proofing), but the name was considered demeaning and the expression 'error proofing' considered less offensive. Poka-yoke can be electrical, mechanical, procedural, visual, human or any other method of preventing an incorrect operation.

Everyone makes mistakes. (As the Darlek said, climbing off the dustbin!) It is part of our human nature. Our aim in our quest for zero defects is to:

Simplify and create a production environment which does not, or cannot produce defects

Let us look at the five stages on the journey to zero defects and the remedies businesses choose along the way:

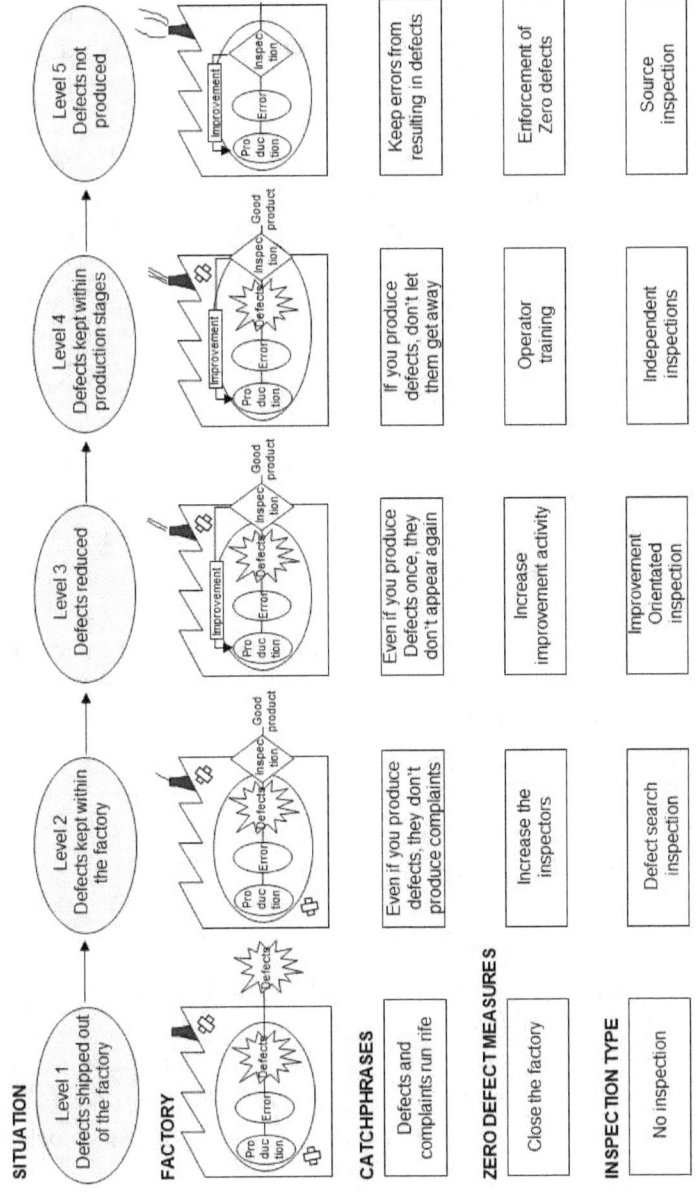

SITUATION

| Level 1 Defects shipped out of the factory | Level 2 Defects kept within the factory | Level 3 Defects reduced | Level 4 Defects kept within production stages | Level 5 Defects not produced |

FACTORY

CATCHPHRASES

| Defects and complaints run rife | Even if you produce defects, they don't produce complaints | Even if you produce Defects once, they don't appear again | If you produce defects, don't let them get away | Keep errors from resulting in defects |

ZERO DEFECT MEASURES

| Close the factory | Increase the inspectors | Increase improvement activity | Operator training | Enforcement of Zero defects |

INSPECTION TYPE

| No inspection | Defect search inspection | Improvement Orientated inspection | Independent inspections | Source inspection |

103

In these days when high quality is expected, indeed demanded in everything we buy and therefore make, Level 1 is totally unacceptable. Even Level 2, where the customer does not receive defects, the high cost of policing quality makes the company uncompetitive. Perhaps now, after achieving stability we are on Level 3 but we are striving to achieve level 5. So let us consider the typical causes of defects.

1. Inappropriate procedures or standards – (Lack of proper planning)
2. Excessive variability in machinery – (Lack of proper maintenance)
3. Damaged or excessively variable materials – (Lack of Inspection)
4. Worn machine parts – (Absence of TPM or tool management)
5. *Human error* – this is what Poka-yoke is designed to minimise and ultimately eliminate.

Human error comes in many forms, let us consider the range:

1. Forgetfulness
2. Misunderstanding
3. Wrong identification
4. Lack of experience
5. Willful (ignoring rules or procedure)
6. Inadvertent or sloppy behavior
7. Slowness to react
8. Lack of standardisation
9. Surprise (unexpected machine operation, etc.)
10. Intentional (sabotage)

In our example, by designing the key to remain with the petrol cap addresses the error of forgetfulness and for new drivers, lack of experience. We may well add sloppy behaviour since driving away with no petrol cap, particularly with a full tank is not a good idea!

By introducing Poka-yoke systems which use sensors or other devices in process equipment to detect errors by the operators or the process itself, we not only achieve 100% Inspection but also quick feedback on the problem. At the extreme, we can design Poka-yoke systems to detect errors and close down the system

There are three distinct methods for using Poka-yoke systems

- Contact method – Detects whether a product makes physical contact
- Fixed value method – A device for counting set numbers
- Motion stop method – If a motion or stop in the process is carried out

1. Physical contact sensing devices

Used in extensively in automated system these work by physical touching the product sending an electric signal when touched. The signal can be used to stop or start a machine of to trigger a warning signal.

Limit switches or micro switches

These are the commonly used physical contact sensing devices. These are used to confirm the presence and position of objects that touch the small lever on the switch.

Touch switch

Similar to a limit switch but activated by a light touch on a thin antenna

Trimetron

This is a needle type gauge that sends a signal when conditions it is measuring are not within the acceptable limits. It is used to open and close gates separating acceptable from unacceptable products.

Go-no go type gauge.

In operations that rely on operators placing components, these are simple jigs that stop the object being wrongly placed. An example is shown below:

Before Mistake-Proofing

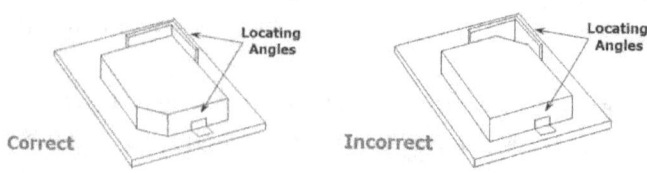

With the jig in its current form, it is possible to place the component wrongly. The answer is to simply add a dowel at a point where it will physically stop the part seating:

After Mistake-Proofing

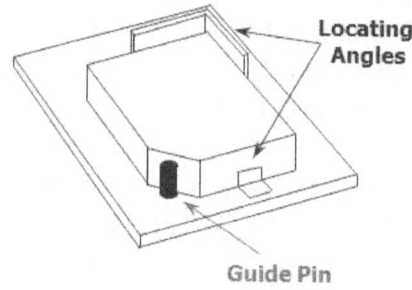

A similar example on a chuck shows how the part is stopped from being wrongly machined:

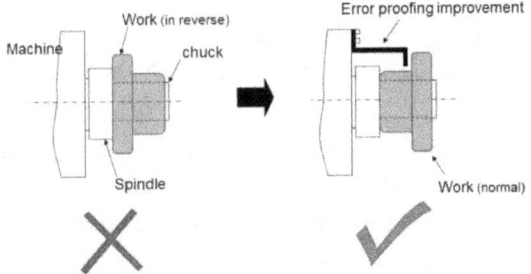

2. Energy sensing devices

Energy sensing devices use energy rather than physical touch to determine whether an error is occurring.

Photoelectric switch

Uses beams of light to detect the proper size or colour, position on a conveyor, the correct supply of parts the small lever on the switch.

Proximity switch

Responds to changes in distance form an object and to changes in the magnetic force, those of us with reversing sensors on our cars are very familiar with these types of sensors.

Beam sensors

These use a beam for detection, for instance for ensuring the correct liquid level in a container.

3. Sensors that detect changes in physical conditions

These detects changes in physical conditions in three categories

Pressure – Pressure changes can be detected by pressure gauges and pressure sensitive switches. They are used for instance to detect interruptions in the oil supply.

Temperature – Through the use of heat activated devices temperature changes can be detected. They are used for instance to detect the surface temperature of dies.

Electrical current – this is used in processes such as welding where it is possible to monitor the weld strength by detecting secondary voltage changes in the weld points.

I am sure by now you have grasped the principle of Poke Yoke and can relate it to your daily life in and outside of the workplace. Let us now look at another technique that creates repeatability in the process itself.

PROCESS CAPABILITY

In my long and varied career, in the 1980s I was the Manufacturing Manager of a company which produced injection moulded containers. These were the kind of containers you might find in a DIY store filled with fence or emulsion paint. They were produced in a low cost, high throughput environment. As the customers often reminded us, people are happy to pay for the contents but not the packaging.

The injection moulding machines dropped out containers every eight or so seconds for the smaller ones, maybe twenty seconds for the large 30L containers. Margins were tight and volume throughput and low levels of scrap were critical if we were to make a profit. Yet, we did, until one day, the managing director appointed a new quality manager. Her first job she explained was to tighten the quality and that translated into tightening the product specs. So, without any knowledge of the injection moulding process and the machinery used to manufacturing containers, she set very tight limits on the product dimensions.

Now, the reasons we were able to make the containers at speed was that the moulds were cooled with chilled water. The plastic entered the mould at 130 degrees and as soon as it was solid enough, it was ejected from the mould and left to finish cooling in the ambient air which, unlike the mould temperature, was not controlled. Two thirds of the cooling was therefore left to chance, which when shrinkage levels are between 2% and 4%, meant quite a wide variation in size. It never mattered to the customer, of course. As long as the lids fit and the container did not leak, they were happy.

I imagine by now you are ahead of me. We went overnight from a profitable high volume container producer to an organisation whose new popular phrase was, "has anybody seen a good one?" None of the production could meet the new specification, and the technicians were tearing their hair out trying fruitlessly to produce to the new specifications. Since QC reported directly to the Managing Director, my pleas to loosen the specs fell on deaf ears. "It's not my problem," was her reply.

Of course it was her problem and I asked to convene a meeting with her and the MD. He was obviously concerned and whilst wanting to support his new manager, could not tolerate a production operation that could no longer produce. "Let us do a process capability study," I suggested, "if we need to improve quality, fine but at least let us determine what these machines are capable of before we tighten the specs."

It was a detailed exercise, but in the end all parties were satisfied. Surprisingly, the new specs were very similar to the old ones!

I think, from my example, you have the gist of process capability and its importance. We are not going to delve too deeply into the subject, this is more for those who wish to progress towards six-sigma. We need an understanding of the concepts, however if we are to understand our next topic, Statistical Process Control (SPC).

Let us start by considering two marksmen. The pattern of their shooting is shown below:

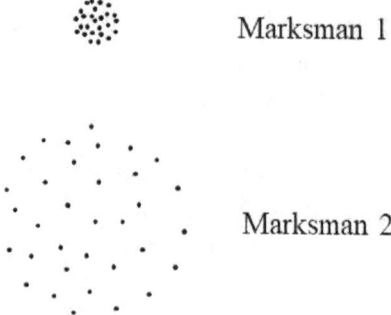

Marksman 1

Marksman 2

So, which of the two marksmen is the most capable? At first sight it is the one at the top. But what are the customer requirements?

Marksman 1   **Marksman 2**   **Target**

In this instance, it is clearly marksman one. But let us look at two different customers. Who is the most capable now?

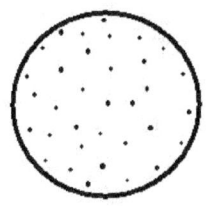

Marksman 1   Marksman 2

So, like our example at the beginning, it really is 'horses for courses'. If you are supplying a high tech electronics company you can expect there to be tight limits on the dimension of your product, if you are providing a container to hold putty, not so stringent as long as the lid fits and it doesn't leak.

A *process capability study* uses data from an initial run of parts to predict whether a manufacturing process can produce parts consistently that meet the required specifications. Think of it as being similar to a forecast. We take some historical data and extrapolate out to the future to answer the question "can I rely on this process to deliver good parts?" More sophisticated customers may require a process capability study as part of a purchase agreement to ensure that the manufacturing processes are capable of consistently producing good parts.

The Basic Concept

When defining the quality requirements of our manufacturing process, our goal is to ensure that the parts produced fall within the Upper and Lower Specification Limits (USL, LSL). A Process Capability Study measures how consistently a manufacturing process can produce parts within specifications.

The basic idea is very simple. You want your manufacturing process:

- To be centred over the Nominal desired by the design engineer, and
- To have a spread narrower than the specification width.

To understand how we achieve this, we need to introduce two new terms; Cp and Cpk

- Cp measures whether the process spread is narrower than the specification width
- Cpk measures both the centring of the process as well as the spread of the process relative to the specification width.

Cp is known as the Customer Process Index and is defined as:

$$Cp = \frac{\text{Specification Width}}{\text{Process Width}}$$

To understand this further, we need to understand the concept of a *normal distribution*. Since I am writing this book in 2020, perhaps the term 'sombrero' springs to mind. A normal distribution looks like this:

$$Cp = (\text{ Upper Spec - Lower Spec }) / 6\sigma$$

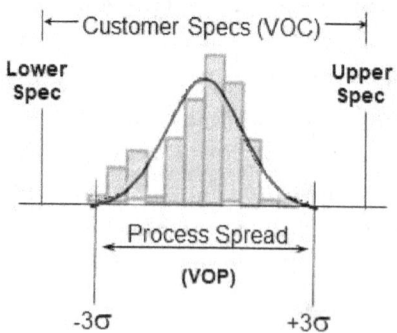

112

We derive this curve from the histogram we introduced in Book One. If you are familiar the term *six sigma*, you will realise where it comes from. As you can see from the diagram, it is the width of the process. A *normal distribution* is the range of values we come across in nature. For example, if I gave you a ruler and asked you to cut 100 pieces of paper exactly 100mm long, how many would exactly be 100mm? "Not many" is the answer; they would probably range between 99.0 and 101mm. If I asked a young child to do it, it might well be between 95mm and 105mm. In both cases, if we plotted the data as a histogram as we learned in Book One, and 'smoothed' it to a curve, it would look like the 'sombrero' shown above.

By calculating Cp, it tells us whether we can meet the customer requirements. If it is equal to 1 it means the process can just about meet the specifications but any shift and it will be outside.

When less than 1, the process cannot meet the specifications and any shift in mean will only make it worse. Many processes operate at this number without knowing it making consistent quality hit and miss. A figure for Cp greater than 1 is what we aspire to, ideally this should be 1.33. At this level we have the comfort of knowing we have some margin for error should conditions change.

Let us now consider Cpk:

$$\text{Cpk} = \frac{\text{Distance to Closest Spec}}{3\sigma}$$

$$= \frac{\text{MIN}\{(\text{USL} - \text{Xbar}\} \text{ or } \{\text{Xbar} - \text{LSL}\})}{3\sigma}$$

Where LSL is the Lower Spec Limit and USL is the Upper Spec Limit, Xbar is the mean and σ is the standard deviation (we will discuss this later). 3σ is half the process spread.

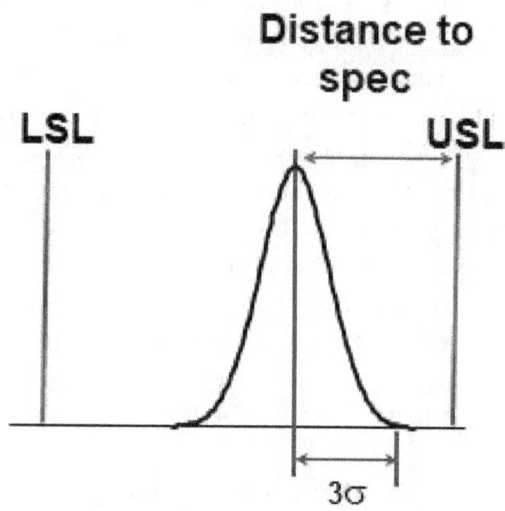

Cpk gives us a measure of whether, even though the process when measured using Cp is capable, the drift from the centre could mean it goes out of specification. Think of it like driving your car into your garage. Normally, if your garage is like mine, it might just fit, the car being slightly narrower than the entrance. However, if you start too near one side, either left or right, and drive straight forward, you are likely to lose a wing mirror! As with Cp, Cpk needs to greater than 1, ideally 1.33.

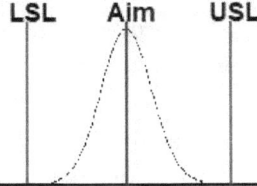

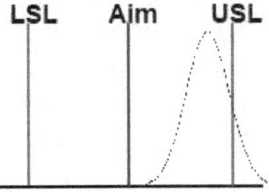

C$_p$ tells you whether process fits within specs.

C$_{pk}$ tells you where your current process is

As you can imagine, there is more to a Process Capability Studies than covered in this short description, but it is necessary to understand the basics if we are to consider our next topic, Statistical Process Control.

STATISTICAL PROCESS CONTROL

Traditional manufacturing relies upon production to make the product and the quality department to inspect it. After-the-event inspection is expensive and wasteful because:

- The product has already been made
- Costly re-work is not always possible

It is much more cost effective to avoid waste by monitoring and analysing the process during manufacture. If we know those factors in the production **process** which affect product quality and **control** them within a range established by the capability study, we need not test the product itself - this is the basis of Statistical Process Control.

Controlling the Process

For a product to be made without rejects, it must be manufactured within specified limits. But a number of factors can prevent this from happening:

Natural Variation:

This is inherent in the manufacturing process and cannot be changed without using a different process or machine.

Common Cause Variation:

These are outside influences that are controllable - temperature, powder quality, speed of manufacturing, skills of the operator are all areas that can be controlled.

An Example of Variation:

Much like us or the child cutting the paper into a given length, a machine cutting straws to length will give an error from straw-to-straw. This is because of the inherent tolerances of the machine – this is known as a **Common Cause Variation,** However, this error is less significant than if we were cutting straws to length using a ruler – this is known as a **Special Cause Variation** (Also called an Assignable Variation). The length of the straws is unpredictable and dependent on our judgement and state of mind as we are cutting.

This raises the question - is my manufacturing process able to manufacture within specification?

In an ideal world, we would measure each and every product that is being made. (When some feature of the product is critical, some companies will measure each part and will use Poka-Yoke techniques to reject the ones out of specification.) In the real world, however, there is not enough time or resource to do this. Instead we measure a sample group of product on a regular basis.

These groups are known as subgroups. The subgroups of data are then plotted onto various charts - in chronological order.

These charts will monitor the variables which are critical to our product. If we were printing plastic lids for instance, these might include:

- lid thickness
- lid colour

- Drying-oven temperatures
- Pre-treatment effectiveness
- Also visual inspection of ink quality and adhesion of the finished product.

These are the data charts we will use to control our process.

Control Charts

We introduced control charts when we considered the 7-Tools of Quality in Book One. (For a resume of the tools, see Appendix I). A Control Chart is tool for decision making that

- distinguishes between common and special cause variation
- uses statistically derived limits based on data from the process so that it,
- helps us recognise when the process **has not** changed when only common cause variations are present
- helps us recognise when the process **has** changed when special cause variations present

Variations, both common cause and special cause variations, occur for a number of reasons, some of which are highlighted in the table below:

Characteristics of Variation

Common Cause Variation	Special Cause Variation
Process operating as usual	Process has "changed"
Part of the system	Unusual
Many, trivial	Few in number, important
Hard to detect and identify	Easy to detect and identify
Predictable over the long run	Unpredictable
Inherent to process	Specific Event
Remains until system is changed	Signals a change
Also called "Chance" Cause	Also called "Assignable" Cause

Special causes indicate an "out-of-control" process

Let us now look at a control chart. In its simplest form it allows us to plot a value of a process variable, critical to the quality of the finished product, on the y-axis against time on the x-axis. Shown on the chart is a line representing the mean value that has been determined from data collected when we prepared our histogram. The chart is divided into six zones, three either side of the mean line. At the extremes are the Upper Control Limit and the Lower Control Limit.

The chart has zones marked A, B and C. These are there to prompt different levels of response similar to the control zones we saw in our Pace chart when we discussed pull. The UCL and LCL come from our earlier discussions and we know the range between them is 6σ where σ is the standard deviation.

118

Control Chart Zones

```
┌ - - - - - - - - - - - - - - - - - - - - - - - -UCL
│          A
├────────────────────────────────────────
│      B
├────────────────────────────────────────
│  C
├──────────────────────────────────── Xbar
│  C
├────────────────────────────────────────
│      B
├────────────────────────────────────────
│          A
└ - - - - - - - - - - - - - - - - - - - - - - LCL
```

If you are not familiar with the term standard deviation, it is calculated from the original data we acquired using the following formula:

$$\sigma = \sqrt{\frac{\sum(X - \overline{X})^2}{n-1}}$$

$\sum$ means "add up"

X = Individual observations

$\overline{X}$ = average

n = number of observations

Calculating Standard Deviation.

Let us say in our process capability study we collect 100 pieces of data. The first thing we do is add them up and divide by 100 to get the average, in the equation it is the X with the bar over it (Xbar). Taking each data point we now take away the average and then square the number (times it by itself). Clearly, any negative numbers will become positive once we square it.

119

Next, we add up all the squared values and divide this by the number of samples less one. In this case, it is 99. Finally we take the square root of this number which gives us our standard deviation. Clearly, the more data we have, the more accurate will be our findings.

What is the significance of this number? It has been shown mathematically that 68.27% of all data will lie within one standard deviation and 99.7% within 3 standard deviations. Hence, when we set the UCL and LCL at this level, we are pretty sure the process is capable of achieving specification.

Control Chart Limits

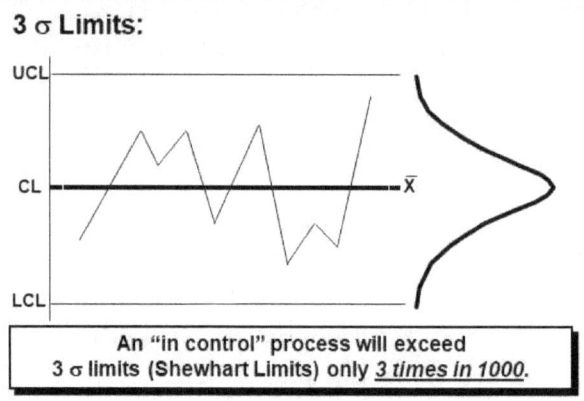

By using this method, the limits calculated from the data in a standard manner means that our limits are the same as the customer's or anyone else analysing the data. They represent common cause variability, in other words, this is the best the best my process can do right now. They also define the "natural process limits" of where our process has been operating and where it will continue to operate unless changed. So far, so good. Let us look at a typical control chart:

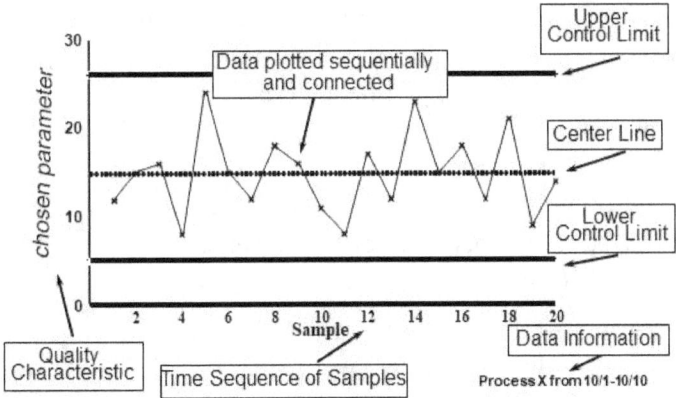

Everything is clearly in control in this chart but what do we do if one of the data points strays outside the upper or lower control limits? This is where we need a set of rules to guide us. Without them, the operator might start adjusting the processing conditions without truly knowing what is causing the deviation.

Decision Rules

We need to decide whether the data is telling us we have moved from common cause to special cause variation. The guidelines to test for Special Causes are:

Rule Number 1. One or more points are beyond the control limit.

Rule Number 2. Nine consecutive points are on one side of the average.

Rule Number 3. Six consecutive points are increasing or decreasing.

Rule Number 4. Fourteen consecutive points are alternating up and down.

Rule Number 5. Two out of three consecutive points are on the same side of the average in Zone A or beyond.

Rule Number 6. Four out of five consecutive points are on the same side of the average in Zone B or beyond.

Rule Number 7. Fifteen points in a row are in Zone C, above and below the average.

Rule Number 8. Eight points in a row are on both sides of the average with none in Zone C.

Wow! That's a list to remember, but of course, it is something we can come back to again and again if we believe our process has moved to special cause variation. Let us look at the first case where Rule Number has been broken:

Rule 1: One or more points beyond the upper or lower control limit(s)

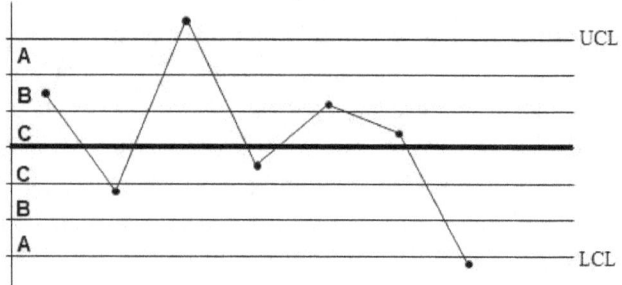

Clearly the process is moving out of control, we need to determine what type of special cause or process change could either of these indicate? Without knowing the operation, it is hard to make a judgement; however the 7-Tools of Quality and other problem solving techniques we learnt in Book One can now come into play.

It is beyond the scope of this book to delve further into the variety of problems that control charts can highlight, it is at the moment sufficient to know the principles behind SPC.

SKILLS FLEXIBILITY

I think it is becoming clearer and clearer as we progress through the book that introducing flow and then pull requires considerably more flexibility from our operators. Rather than the specialisation of Henry Ford's wheel nut tightener, our lean operator has to have process skills, problem solving skills and, in the case of lightening quick changeovers, some mechanical aptitudes. Furthermore, if we are to operate in production cells, the chances are that the operator will need to know how to set up and operate each unique machine in the cell. So how do we manage this, how can we ensure that we have the human resources to operate with flow and pull? The answer is of course, to create a skills matrix. We touched on this briefly in Book One when we considered Andon systems, for in true lean form, we need to make the Skills Matrix part of Visual Factory rather than something hidden in a file in the training department. So, let us look at what constitutes one:

Skills Matrix

- It is an integral part of our Visual Management System (VMS).
- It is a simple visual tool to aid in the management, control & monitoring of skill levels.
- It displays all tasks & skills required to work in an area or team.
- It displays all current team members.
- For each team member it displays current competency/ability levels for each task.
- It is a simple tool to aid resource planning.

Let us look at a sample skills matrix:

Team/Area:				Team/Area Leader:			Date:
Skills/Tasks Name							
Skill Level Key							
	Un-Trained	Learner	Practitioner	Developer	Coach		

The first thing to note is how easy it is to read and decipher. There are five levels of skill identified:

- *Un-trained:* As the name infers, this person has no experience of the skill required, the task to be undertaken, and relevant work instructions
- *Learner:* This person is being taught the skill/task/work instructions.
- *Practitioner:* This operator can carry out the skill/task/work instructions,
  - Safely
  - To the correct quality standards, first time
  - Without assistance
  - Can operate up to 1.5 times the standard cycle time. (i.e. is still not up to speed)
- *Developer:* This person can actually improve the skill/task/work instructions,
  - Safely
  - To the correct quality standards, first time
  - Work to the standard cycle time

- *Coach:* Someone who has the skill level of a Developer, but can train & develop others in carrying out the skill/task/work instructions
  - o Safely
  - o To the correct quality standards, first time
  - o Work to the standard cycle time

This person is likely a work-based trainer and will be made available to help bring on less skilled operators.

The Skills Matrix is used:
- To establish all skills required in an area or team
- To visually share information.
- To quickly identify current available skills and future requirements.
- To examine where our strengths & weaknesses are.
- As a day to day planning tool to use skills where they are most needed.
- As a planning tool to organise adequate cover for holiday & sickness.
- If done fairly, to keep employees motivated & reduce boredom.
- It highlights training needs for our most important resource....our employees.
- To monitor & control training effectiveness.
- To increase flexibility by allowing people to master a broad range of skills.
- To drive improvements as part of a Visual Management System (VMS).
- To increase the effectiveness of an area and the entire business.

The Skills Matrix plays a major part in the preparation of the Company's training plan. The training plan should be aligned to the needs and demands of the business. If the business operates an employee appraisal scheme, it can form the basis of the annual review. It also serves to monitor & control training effectiveness and to increase flexibility by allowing people to master a broad range of skills.

Let us look at an example of an actual skills matrix:

•Team/Area:	•Station 1				•Team/Area Leader:		•Date:
•Skills/Tasks •Name	•Milling	•Drilling	•Deburring	•Grinding	•Painting	•Riveting	•Name •Score
•Tom							11 / 24
•Dick							14 / 24
•Harry							10 / 24
•Skills/Task •Score	3 / 12	8 / 12	8 / 12	5 / 12	4 / 12	7 / 12	35 / 72
•Skill Level Key	•Un-Trained	•Learner	•Practitioner	•Developer	•Coach		

The matrix is easily read and highlights not only strengths but deficiencies. It is good practice to have at least two team members up to coach level in each individual skill. This gives the team leader and shift leader flexibility when covering holidays and sickness. The skills matrix should be a live document and be handwritten and kept visible in the work area. By all means use a pre-printed blank.

Ownership of the Skills Matrix

The skills matrix should be owned by the leader of the particular section where it is displayed. There should be a formal assessment by either the team leader or the work-based trainer before a team member is considered to have the appropriate skill. The skill should be quantifiable where possible, for example the ability to operate the machine at the product Takt time. Once the skill has been assessed and confirmed it can be added to the matrix by the assessor.

There may be grey areas when deciding about skill levels. What happens if someone does not do a particular job for a while? Do they keep the same skill level? The solution might be:

- Freeze their skill level pending an assessment of the skill in question.
- Indicate on the skills matrix that the person needs to be assessed beside the skill in question.
- Agree a date with the employee as to when they are going to be re-assessed.
- Update the skills matrix after the assessment.

Skills flexibility not only benefits the company by optimising machine and labour utilisation, training is also a highly powerful motivator. When we discuss leadership in Book Three we will realise there are very few ways to motivate people. Standing in front of them and telling them you are going to motivate them is a sure way of failure; try it and watch their arms fold in unison! Motivation is something we do to ourselves not something which is done to us. Offering training in new skills to an individual motivates at a number of levels. It confirms they are valued as team members, it makes them feel more secure (companies tend not to train people and then dismiss them) and it gives them an opportunity to show off their own skills and attributes. As we have seen throughout the book, lean is embraced readily by employees because it makes their job so much more interesting.

MODULE 6

VALUE STREAM MAPPING

Value Stream Mapping (VSM) is the start of any lean transformation. To understand VSM we must first clarify the difference between Flow and Pull, the two key lessons of this book. If we create true Flow we actually don't need Pull. The product is literally manufactured in one piece and in theory, WIP in the form of Kanbans are not required. I used to supply the cases for Parker Pens and I was amazed to watch a ballpoint pen assembled, automatically, in one operation, from start to finish including test writing of each pen. Most manufacturing processes, however, do not lend themselves to pure-one piece flow. The simple rule is:

Where you can flow you flow,
Where you can't flow, you Pull!

Let us begin with a definition:
A Value Stream Map is a diagram of all actions (both value added and non-value added) required to take a product through from raw material to the customers warehouse

As you might expect, the methodology we are about to look at has its beginnings with Taiichi Ohno at Toyota in the 1940s who wanted a tool to represent the process visually. It was known at that time as a Material and Information Flow Diagram. However, it was two Americans from the Lean Enterprise Institute, Mike Rother and John Shook who brought the principles to the attention of the world in 1999 when they wrote the book, *Learning to See.* This introduced us to the term Value Stream Mapping.

I make no attempt in this book to go into the depths of this subject to make you an expert - this was never the intention of the Lean Made Simple series. What I hope, however, is that you will learn enough to review your own process and create two Value Stream Maps. Remember, practice makes perfect and training is most effective when you apply what you have learned within 24 hours.

As you might expect, VSM starts with a map, known as a Current State Map and finishes with a map, the Future State Map. Our focus throughout is on Value and its nemesis, Waste. The whole idea is to make the analysis visual so let us start by looking at the symbols we will use to map the value stream:

MATERIAL FLOW ITEMS

Manufacturing Process (Press)	Used to depict a process, not for individual process steps but for a general process. Usually placed anywhere the flow stops.
Manufacturing Process Shared (Heat Treat)	Used for a process that is shared for many different products. An example would be Heat Treat.
Outside Sources (XYZ Corp)	Used to represent either a Customer or a Supplier
Data Box (C/T=25, C/O=40, 2 Shifts, Taft= 5)	The data box is used to put information about a process, customer etc. It should contain pertinent data. What goes in it is up to you, use it flexibly but get the core numbers.

Truck Shipment (2X per week)	Shows material shipment by truck. Include the frequency or any other important info. inside
PUSH Arrow	Use to show material pushed by a schedule.
Finished Goods to Customer	Use to show finished goods material to a customer or supplied material from a supplier.
First-In-First-Out Sequence Flow (FIFO)	Shows material sequenced first-in-first out where the sequence was initiated upstream based on pull.
Supermarket	Symbol illustrates a supermarket where material quantities are calculated based on demand and lead time. The material movement is controlled using signals or kanban.
Physical Pull	Describes removal of material from an upstream process by a downstream process based on need.

INFORMATION FLOW ITEMS

Manual Info Flow	Shows information flow achieved through manual movement of paper schedules or orders.
Electronic Info Flow	Shows electronic information flow by computers. EDI or E-Mail are examples.
Schedule (Weekly Schedule)	The schedule box is used to show the use of schedules in the information flow.
Load Leveling (OXOX)	Represents a leveling tool that mixes models. (Heijunka box or similar tool)
Withdrawal Kanban	A kanban card authorizing removal of parts or material from store.
Production Kanban	A kanban card authorizing production based on use in a downstream process.
Signal Kanban	Represents a kanban symbol used to authorize production. Multiple cards are not used and the signal can be made in many different ways.
Kanban Post	A collection point for kanban cards to be placed.

Sequenced Pull Ball	Represents a leveling and sequencing tool to control the sequence of production.
Go See Production Scheduling	Scheduling achieved by physically looking to see what a downstream process needs. Not controlled by kanban or calculation.
GENERAL ICONS	
Kaizen Lightening Burst	Used on Future State maps to show where Kaizen activity is needed to execute the future plan.
Buffer or Safety Stock	Represents safety stock in place to cover for inconsistencies, quality etc.

Whilst we display the information visually, we still need to collect data about our products and process. Rather than looking plant wide, we start by looking at a product or a product family.

Selecting a Product Family

The key is not to try and understand and map everything that is happening in our factory. Instead, we map the value stream for one product family at a time.

A product family is a group of products that pass over the same or similar process steps and equipment. We define our product families by looking at processes at the downstream end of the flow, nearest to shipping and hence the customer. Upstream fabrication processes, like stamping, injection moulding, or cutting, often feed parts to a variety of downstream processes, and thus may serve several product families.

If you have already organised your assembly operations into 'product focused cells', or if your assembly is paced by a moving conveyor, probably everything that goes through one cell could be considered a product family. If you have not yet put your operations into cells, then value stream mapping, beginning with the product family matrix, is just what you need.

When the variety of products and assembly-type processes is complicated, a table of products and processes like this can help clarify product families.

		PROCESS AND EQUIPMENT STEPS					
		Weld 1	Weld 2	Assy. 1	Assy. 2	Polish	Pack
PRODUCTS	Left	X	X		X	X	X
	Right	X	X		X	X	X
	Upper	X	X		X		X
	Lower	X		X			
	Inner skin	X			X		
	Outer skin	X	X	X	X		

The sequence of invents is, we select the product group, we map the Current State (CSM) and we develop the Future State Map (FSM). It is important to complete the CSM and the FSM within a short period of time otherwise we will lose too much information. It is important to understand this because whilst collecting data for the current state map we will already be thinking about how we can make improvements. By leaving the future state mapping until sometime in the future, all these ideas will be lost.

To complete the CSM we need to understand how the shop floor currently operates. To do this we collect material and information flows. Using the icons above, we start with the "door to door" flow. Our natural tendency, when showing someone our operation for the first time, is to start from the raw material area. Since we are looking at our operation now in terms of 'pull', we should instead start from the despatch end of the process.

We need to walk the flow and get actuals times that the process is operating to, not the standard times. We use a decent sized sheet of paper and draw the process by hand, with a pencil to make it easier to make alterations. This will be our foundation for the future state. Whilst we are doing this and observing value and waste, at the same time we should start to develop a plan, otherwise all our work will be to no avail.

CURRENT STATE MAP (CSM)

THE 7 STEPS to creating the CURRENT STATE MAP are as follows:

1. Establish Customer Requirements
2. Draw Process Steps
3. Collect Process Data
4. Identify Inventory
5. Determine Material Flow: Supplier to Manufacturer & Manufacturer to Customer
6. Detail Information Driving Flow & Internal Material Flow
7. Calculate Manufacturing Lead Time and Processing Time

Let us look at developing a Current State Map for a family of Widgets.

STEP 1: The **Customer Requirements** are:
- 320 pieces per month
- 200 per month of Type A1 Widgets

- 120 per month of Type A2 Widgets
- Customer plant operates on 2 shifts
- Palletised returnable tray packaging with 1 widget in a tray and up to 10 tray's on a pallet. The customer orders in multiples of trays
- One daily shipment to the assembly plant by Truck

STEP 2: **Draw the Process Steps**

Our *Work Time* is:

- 20 days in a month
- Two shift operation in all production departments
- Eight (8) hours every shift, with overtime if necessary
- Two 10-minute breaks during each shift
- Manual processes stop during the breaks
- Unpaid lunch - all processes continue

We draw the customer requirements on our sheet of paper at the right hand side and then add the process boxes. To manufacture a widget we begin with a stamping process which then sends the production to a spot-weld and so on until the product is ready for shipping.

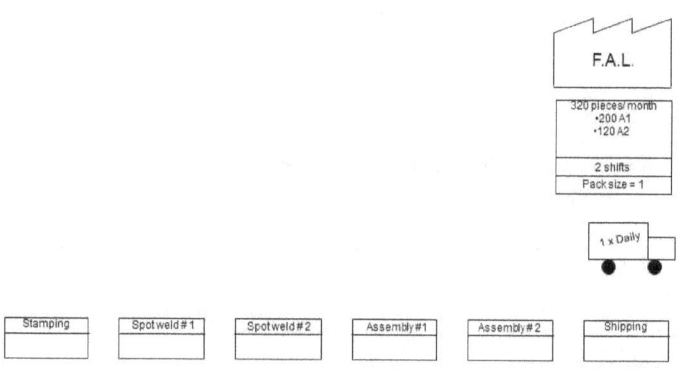

134

STEP 3: **Collect Process Data**

We now need to add the information that we have observed about each stage of manufacture. Let us consider the stamping process:

STAMPING: (The press makes parts for many other products)
- 1 operator to run press
- Automated 10 ton press with coil (automated material feed)
- Cycle Time: 1 minute (60 pieces a hour)
- Changeover time: 1 Hour (Good piece to good piece)
- Machine reliability: 85%
- Observed inventory:
  - ○ 5 Days of coils before stamping
  - ○ 150 pieces of Type A1 finished stampings
  - ○ 24 pieces of Type A2 finished stampings

We add these into the boxes representing the Stamping stage on the CSM:

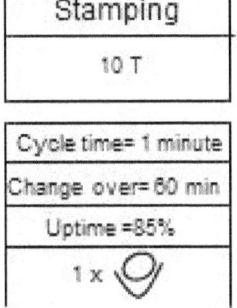

We continue for each stage of the process using the information shown on the next page:

Production Processes: All processes occur in the following order and each piece goes through all processes

1.) STAMPING
 (The press makes parts for many products)
•1 operator to run press
•Automated 10 ton press with coil(automated material feed)
•Cycle Time: 1 minute (60 pieces a hour)
•Changeover time: 1 Hour (Good piece to good piece)
•Machine reliability: 85%
•Observed inventory:
  5 Days of coils before stamping
  150 pieces of Type A1 finished stampings
  24 pieces of Type A2 finished stampings

2.) SPOT WELD WORKSTATION 1
 (Dedicated to this product family)
•Manual process with 1 operator
•Cycle Time: 39 minutes
•Changeover time: 10 minutes (Fixture change)
•Reliability: 100%
•Observed inventory:
  36 pieces of Type A1 after process:
  6 pieces of Type A2

3.) SPOT WELD WORKSTATION 2
 (Dedicated to this product family)
•Manual process with 1 operator
•Cycle Time: 46 minutes
•Changeover time: 10 minutes (Fixture change)
•Reliability: 80%

3.) CONT'D
•Observed inventory:
  6 pieces of type A1 before spot weld 2; 16 pieces
  of Type A1 after spot weld; 18 pieces of Type A2.

4.) ASSEMBLY WORKSTATION 1
 (Dedicated to this product family)
•Manual process with 1 operator
•Cycle Time: 62 minutes
•Changeover time: non
•Reliability: 100%
•Observed inventory:
  42 pieces of Type A1
  6 pieces of Type A2

5.) ASSEMBLY WORKSTATION 2
 (Dedicated to this product family)
•Manual process with 1 operator
•Cycle Time: 40 minutes
•Changeover time: none
•Reliability: 100%
•Observed finished goods inventory in warehouse:
  27 pieces of Type A1
  14 pieces of Type A2

6.) SHIPPING DEPARTMENT
Removes parts form finished goods warehouse and
stages them for Truck shipment to customer

STEP 4: **Identify Inventory**

Next we take the inventory between the processes, which I have shown as a triangle with an 'I' inside and add this to the CSM:

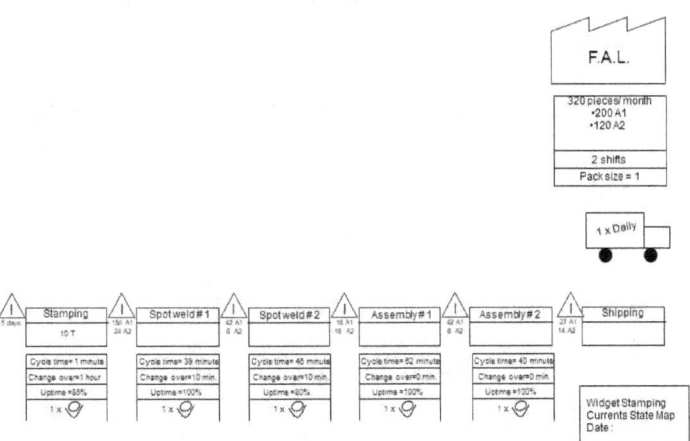

STEP 5: **Determine Material Flow**:

In this step we add an icon for the customer, add an icon for a lorry and arrows to show the flow of material:

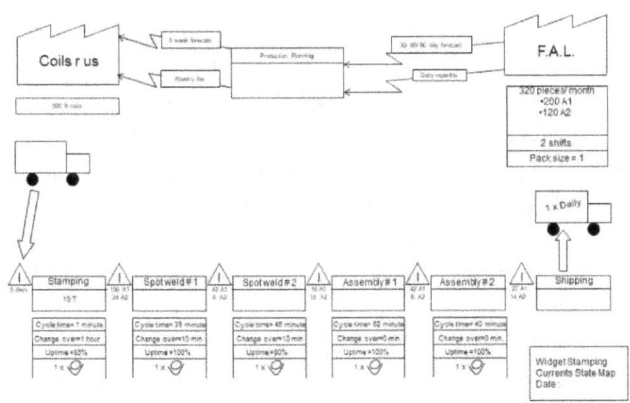

Step 6: **Detail Information Driving Flow & Internal Material Flow**

This is the stage where we add the information which is used to schedule each process. We mark up the CSM with a pair of glasses to remind ourselves that we need to go and check the progress of the material on the shop floor!

STEP 7: **Calculate Manufacturing Lead Time and Processing Time**

The method for turning inventory into time is by using the following formulas:

$$\frac{\text{Number of products}}{\text{Takt}} = \text{time delay for inventory}$$

$$\frac{\text{Inventory quantity}}{\text{Customer demand}} = \text{Number of stock days}$$

For example, the stock between stamping and Number 1 Spot-weld is 174 units

Customer demand is 320 units per month over 20 days delivery = 16 units per day

$$\text{Number of Stock Days} = 174/16 = 10.875 \text{ rounded up}$$
$$= 10.9 \text{ days}$$

Congratulations! We now have our Current State Map. Already we can see areas of waste we can focus on.

CURRENT STATE MAP

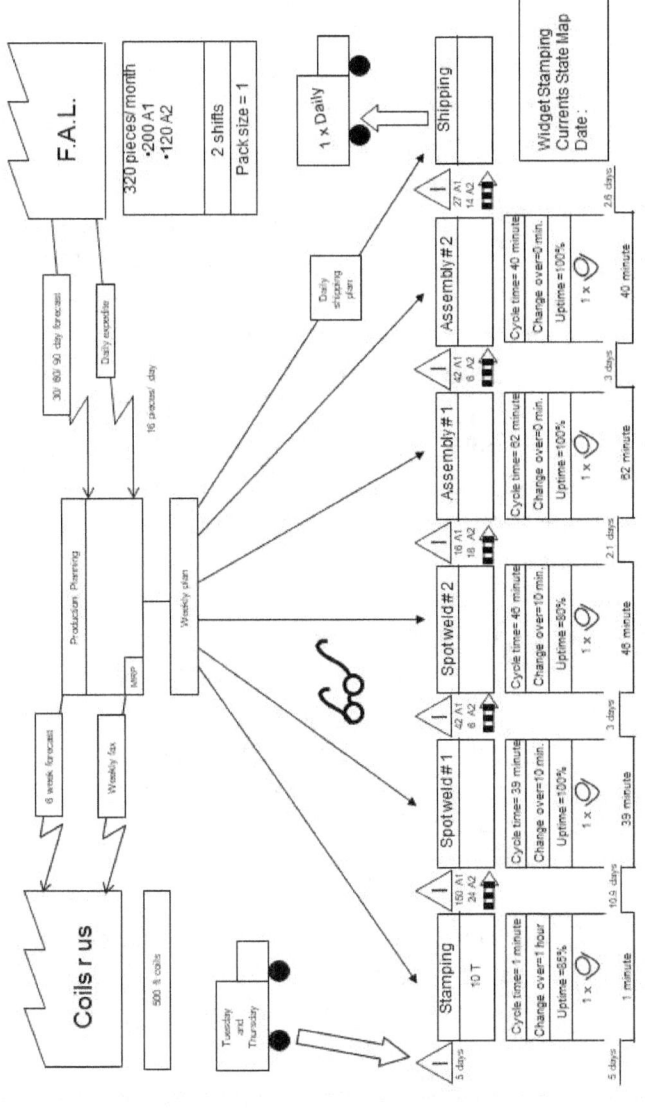

320 pieces/month
•200 A1
•120 A2

2 shifts

Pack size = 1

F.A.L.

Coils r us

500 ft coils

Production Planning

MRP

Weekly plan

300/ 60/ 90 day forecast

Daily expedite

6 week forecast

Weekly fax

16 pieces/ day

Daily shipping plan

1 x Daily

Shipping

Widget Stamping
Currents State Map
Date :

Tuesday and Thursday

Stamping	Spotweld # 1	Spotweld # 2	Assembly# 1	Assembly # 2
10 T				
Cycle time= 1 minute	Cycle time= 39 minute	Cycle time= 46 minute	Cycle time= 82 minute	Cycle time= 40 minute
Change over= 1 hour	Change over=10 min.	Change over=10 min.	Change over=0 min.	Change over=0 min.
Uptime=85%	Uptime=100%	Uptime=80%	Uptime=100%	Uptime=100%
1 x	1 x	1 x	1 x	1 x
1 minute	39 minute	46 minute	82 minute	40 minute

5 days

5 days

150 A1
24 A2

10.9 days

42 A1
6 A2

3 days

16 A1
18 A2

2.1 days

42 A1
6 A2

3 days

27 A1
14 A2

2.6 days

139

Before we move on to design the Final State Map let us remind ourselves of the key lean techniques we are to use. We need to:

- Calculate **Takt** time
- Identify **Value**
- Eliminate **Waste**
  - Overproduction, Waiting, Transportation, Over-processing, Inventory, Scrap/Rework, Motion.
- Introduce **Flow** (where we can) and
- **Pull** (where we can't).

Overproduction is the worst waste of all the wastes. Overproduction can cause all the other wastes; by manufacturing product too early for 'Just-in-Case' this can lead to too much inventory. The interruption of the smooth flow of goods and services can create unnecessary transportation and waiting time. Let us begin by calculating the Takt Time:

The Available Time is calculated as follows:

Takt Time = Total Time Available*
          Total Customer Demand

Customer Demand = 320 pieces per month over 20 days
             = 16 pieces per day.
*Total Time Available = 8.5 hour shift less 30 minutes lunch less 10 minutes team time
             = 450 mins per shift x 2 = 900 minutes

Takt Time = 900/16 = 56.25 minutes

Now that we have the Takt time for this family of products and the individual cycle times for each resource, we can look

140

where flow and pull are possible. To do this we draw the individual cycle times on a bar chart with the Takt time added:

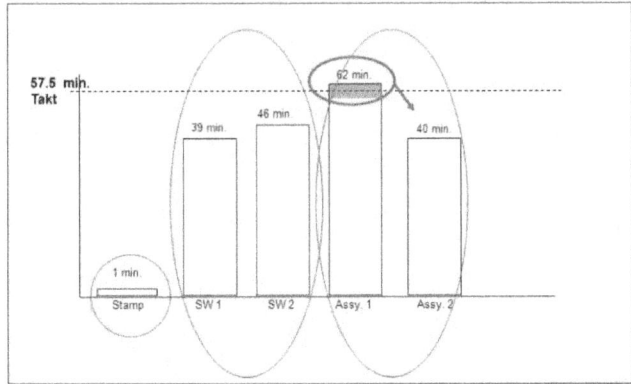

The first thing we can see is that, since Stamping is used by many other products, it cannot become part of a dedicated flow.

The next thing that is evident is that the cycle time of the two spot welding operations are similar and offers us an opportunity to introduce flow.

The next observation is that, since assembly is a manual operation, we should be able to achieve continuous flow between the two assembly operations.

Let us look how a pull system might look:

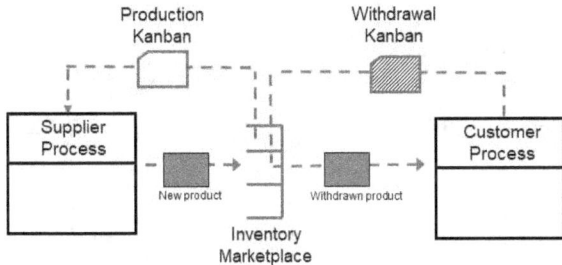

Since information flow is now by Kanbans and the WIP stores can be replaced by supermarkets, we can add this to our Current State Map, together with the flow we have created in Assembly.

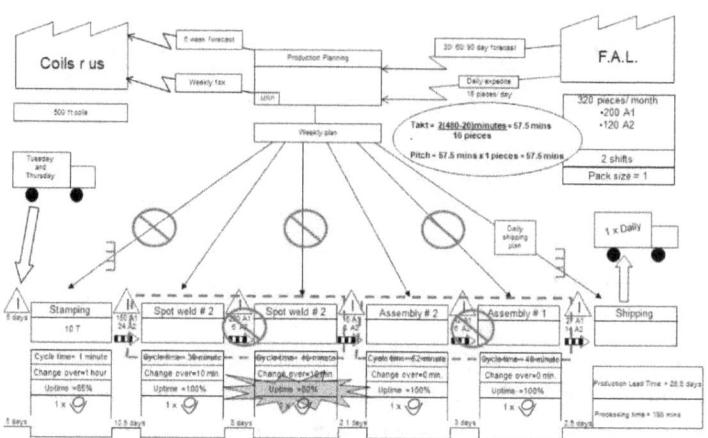

Already we have saved 3.8 days lead time. Before we break out the Champaign, let us look further into the options we have identified starting with the labour content.

$$\text{Ideal Number of Operators} = \frac{\text{Total Work Content}}{\text{Takt Time}}$$

Total work content = 1 + 39 + 46 +62 + 40 = 188.
Since the first operation is shared and cannot be dedicated, we can take this out of our calculation.

$$\text{Ideal Number of Operator} = \frac{187 \text{ minutes of work}}{57.5 \text{ minutes Takt}}$$

= 3.25 operators rounded up to 4 in total

Our target should be to reduce this to three operators. Let us revisit our chart and see if there are more opportunities for flow. We know the Takt time is 57.5 minutes. If we can load each operation in the process to a figure just short of this, we could further even out the work. 56 minutes is too close to the Takt time, if we chose 55 minutes it would give us nearly 5% margin. Let us once again at the opportunities for flow and look how this might be done:

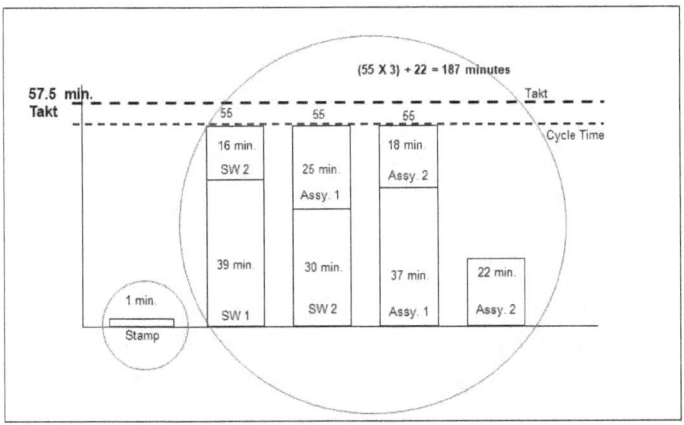

We have moved work from one process to the earlier process to better utilize the operator. Remember, we are not trying to run each process flat out as previously, but produce to the Takt time. This would be achievable with a good kaizen (CI) process in the company. This should be possible because we spend a lot of time fetching bits.

By choosing a cycle time lower than the Takt time;

- It allows some time for inefficiency because cycling to Takt can be risky in a production system that shows a degree of instability.
- It can be the start of the kaizen cycle for the reduction of man hours
- It can be an indication of the flexibility in the system.

Let us redraw the Value Stream Map once again showing the new thinking, this time showing a Kaizen Burst at spot weld number 2 to reduce the changeover time.

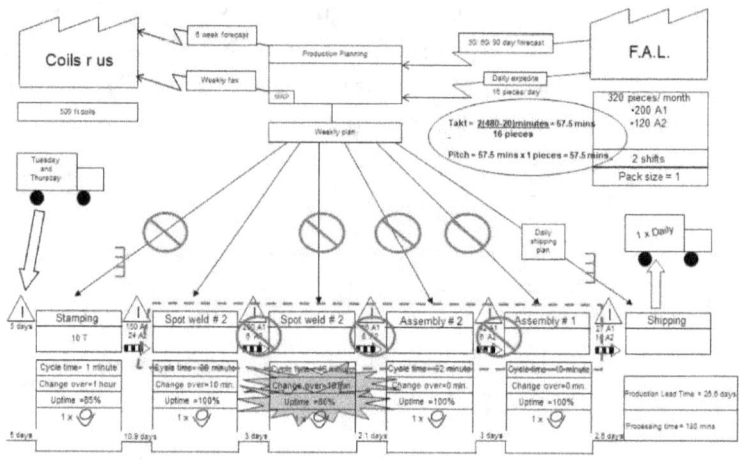

We are now really making progress, although our CSM is looking rather messy. Let us tidy it up in a way that represents the Pull we have created, adding the supermarkets at either end.

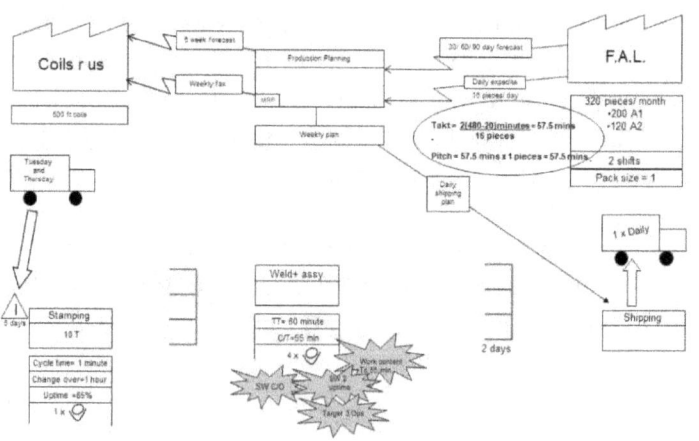

Since we are making radical changes to the method of working, I have decided to add 2 days stock of finished goods. Also, Stamping is not part of the flow so we need a supermarket to schedule it. To service the new layout let us introduce 'material handlers' to service the line and add value more frequently. We now can introduce production and withdrawal Kanban, as per our diagram earlier.

Since the stamping process is shared by many other products, it is not practical for the stamping process to produce exactly to Kanban requirements. The reason for this is simple; the stamping process needs one hour to change over.

We can overcome this by managing the optimum inventory through minimum batch quantities to achieve flow for all the processes that it supplies. We have to determine the batch size by introducing the concept of Every Part Every Interval (EPEI). This is a levelling technique we will consider in more detail in Book Three. As its name suggests we want to produce every part every day. This then determines that the supermarket stock level will be 1 day but, to be on the safe side we will add another half a days' worth. This is an insurance against problems that may occur until we are fully confidant in the system. The batch Kanban tells stamping to change over for the next product.

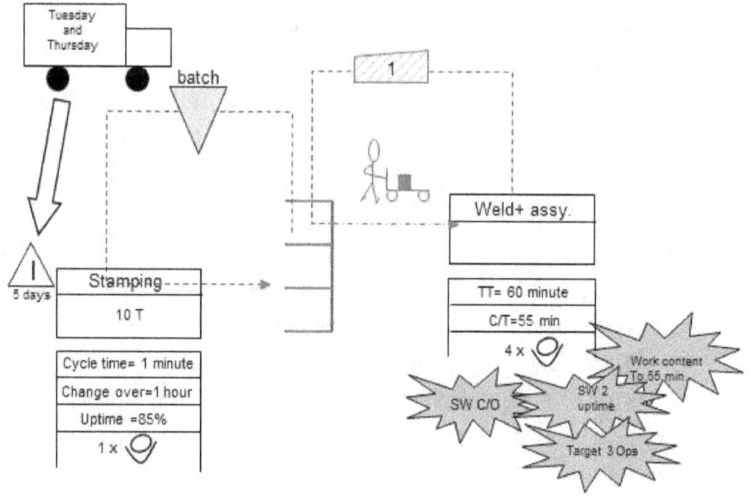

We now write 1½ days under the stamping supermarket. Shipping takes 46 trays out of the assembly Supermarket each day and prepares or stages them ready for dispatch. This releases 46 Kanbans to assembly. Sending 46 Kanbans back to assembly, however, is not very good. When we get to this level of Kanbans we are dangerously close to a mass production environment again.

So, what can we do? Of course we need to level the production onto the assembly process in a sequenced way. The assembly operation actually only needs one Kanban at a time if we sequence them correctly. We know that for our widget product group we have to produce 150 pieces per day of Type A widgets and 40 pieces a day of type 2 widgets. We need to introduce a Load Levelling Box where we can display the Kanbans in the right sequence to achieve the customer requirement.

146

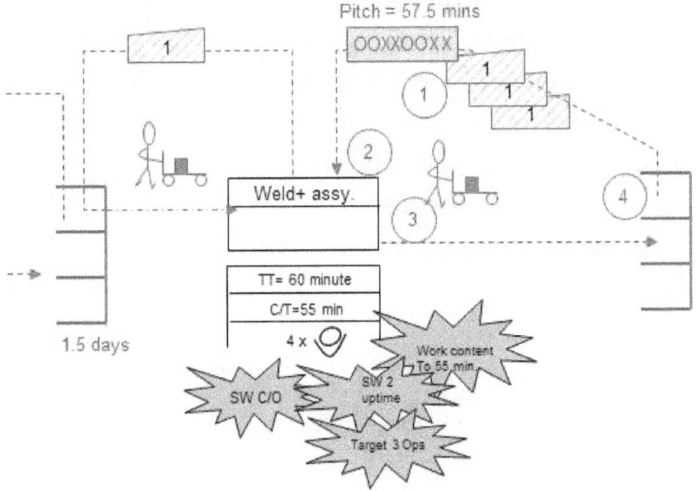

The load levelling box shows the Kanbans awaiting processing in an order that will meet the daily requirements for the two variants of widgets.   The material handler needs to operate in the following manner:

1. Pick up next Kanban at the leveling box
2. Take the Kanban to assembly to tell them what the next 1 minutes production will be
3. Pick up the latest 1 minutes-worth of completed work (1 pitch worth)
4. Takes the finished pieces to the supermarket

How long do you think it will it be before we become aware of any production problems? One 1 minute! Contrast this to the previous method of working where defective parts could be hanging around for days before we came across them.   One minute is now our management time frame.

The next supermarket, prior to shipping, has to be decided by negotiating with the supplier. We need to agree a change to daily deliveries and hold one and a half days to stock, as we have before the stamping process, as a minimum. It is prudent to suggest to the customer that we hold a further ½ days-worth of comfort stock bringing the stock before dispatch to two days.

Finally we add the new timeline and we have our Final State Map.

FUTURE STATE MAP

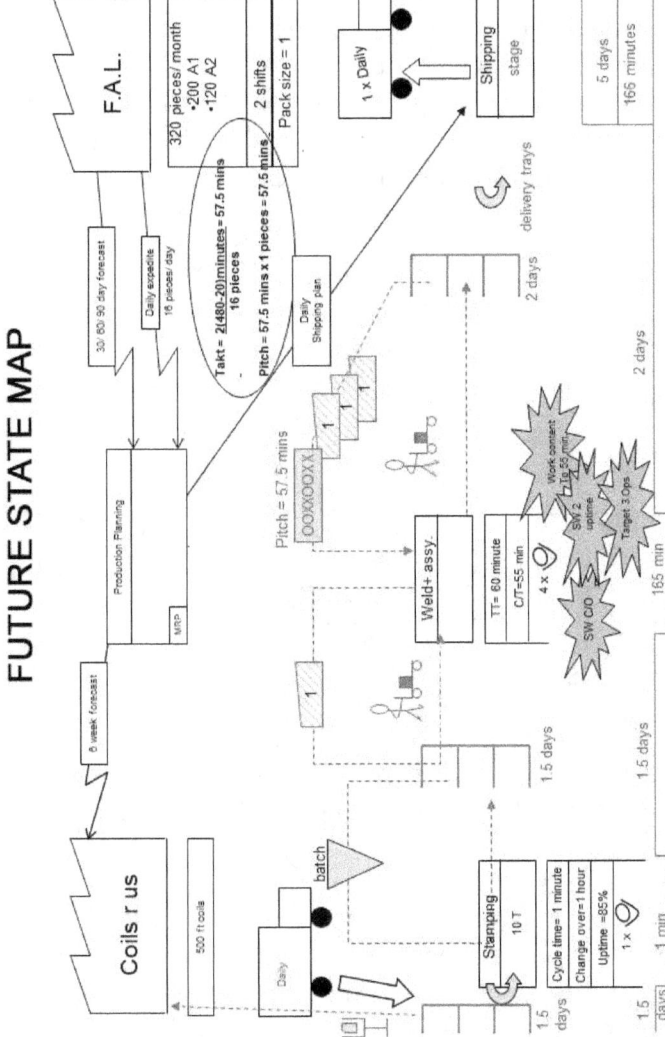

Let us compare the difference between the Current State and the Final State. Lead time to the customer has been reduced from 26 days to 5 days. Manufacturing time from 190 minutes to 166 minutes and the operators employed in the process have been reduced from 5 to 4. Of course, the Final State is not set in stone; it too can be improved together with the supply process. Think of the map as containing three distinct areas:

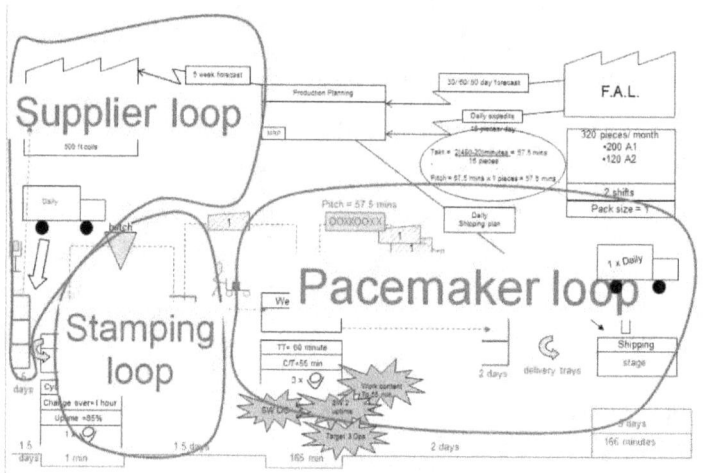

There are further benefits to be gained in looking at the stamping loop and the supplier loop, but that's for another day.

SEMICONDUCTOR PLANT EXAMPLE

The above exercise shows the steps we can take to add flow and pull into an operation and considerably reduce stock and customer lead times. I offer next an example from my own career, although as you might expect, I have to sign a confidentiality agreement with a client whenever I am invited to work with them so cannot show business-sensitive information..

I am going to show you, briefly, parts of the before' and 'after' from an assignment to show the gist of what can be achieved.

I said earlier, we were working with a project team to add lean into the final operations area of the FAB. Below is part of the Current State Map.

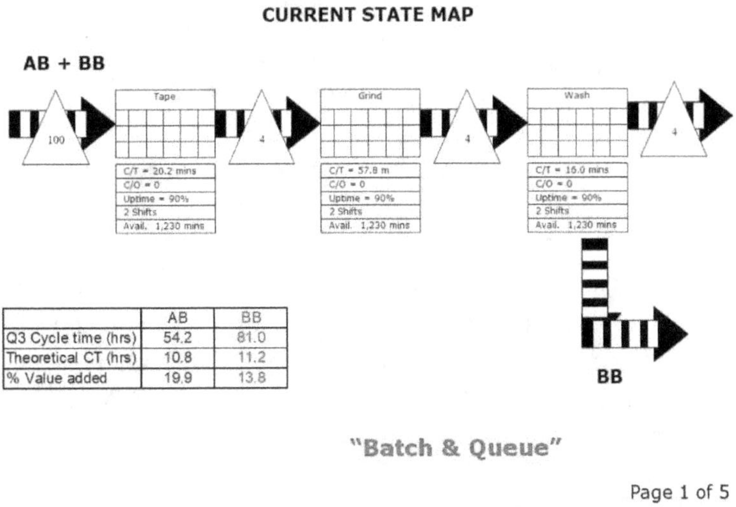

CURRENT STATE MAP

AB + BB

	AB	BB
Q3 Cycle time (hrs)	54.2	81.0
Theoretical CT (hrs)	10.8	11.2
% Value added	19.9	13.8

BB

"Batch & Queue"

Page 1 of 5

Names and information has been changed to maintain confidentiality. By applying the same methodology, the Future State Map for the same area is shown below:

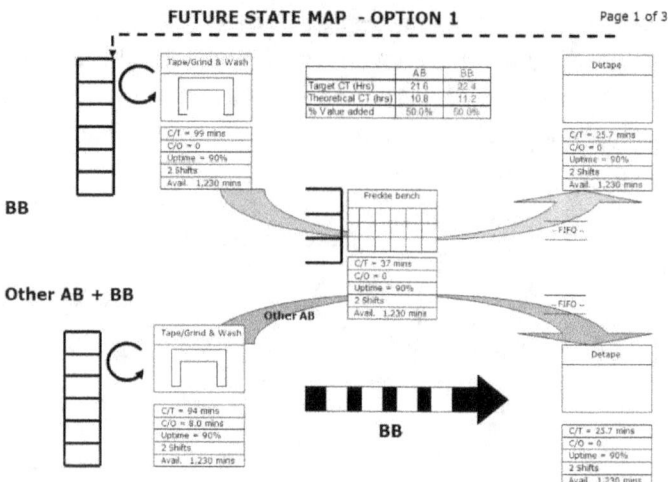

The new way of working reduced the number of operators required for the section from 11 to 8 with a considerable reduction of lead time and increase in tool utilisation.

Remember, it all starts with a Current State Map. Below I summarise the sequence of events that we followed to prepare the CSM

- Draw **customer, supplier** & **production** control icons
- Enter **customer requirements** per month, per day
- **Calculate daily production** & **container** requirements
- Draw **outbound shipping icon** & truck with delivery frequency
- Draw **inbound shipping icon**, truck and delivery frequency
- **Draw boxes for each process in sequence, left to right**
- Add **data boxes** below the process boxes & Timeline for Value-Added & Non-Value Added
- Add **communication arrows** and note the methods and frequencies
- Obtain **process attributes** and add to the data boxes. It is best to observe all times directly

- Add **operator symbols** and numbers
- Add **inventory locations** and **levels** in production units
- Add **push, pull and FIFO** icons
- Add any other useful information
- Add **working hours** - use net available hours planned or scheduled
- Calculate **Lead Times** and place on the timeline
- Calculate **Total Cycle Time** and **Lead Time**

They say a journey of one thousand miles starts with one step. I strongly encourage you to look at a process in your operation and draw the CSM following the guidelines I have given you. Once complete, use the lean tools we have learnt to draw the FSM and I assure you, you will be amazed at the opportunities it will present. Start small, become comfortable with the process and you will slowly become proficient.

The Chinese, thousands of years ago, also encouraged innovation and had a proverb I'd like to share with you. It says, "be not afraid of growing slowly, be afraid only of standing still.'

CREATING PULL AND FLOW - SUMMARY

So, here we are at the end of the second book. I said in the introduction that introducing pull and flow requires courage; it requires working in a completely different way. For those of you in the front line of manufacturing, doing the day job can seem difficult enough without the added complication of introducing what may seem a radical change to your operation.

I was once invited into an organisation as part of a team of consultants to help put in place a Management Control System. It was a bakery, primarily manufacturing cakes. Their biggest and most important customer was Marks and Spencer, not an easy task master at that or indeed, any time. To call the operation 'chaos' would be to credit them with more organisational skills than they possessed. They spent the entire day 'chasing wagons', frantically trying to ensure they met that day's delivery schedule. Everyone from the team leader on the line to the production director was frantically progress-chasing the products needed for the delivery. The production director, who should have been stepping back to ask the question, 'how do we get out of this mess?', was the worst offender, up to his eyeballs in firefighting whilst those whose job it was stood aside and rolled their eyeballs. It reminded me of the saying, 'when you are up to your neck in crocodiles, try to remember that you are here to drain the swamp!' (Needless to say, the company went out of business a few years after I left).

Having said that, (it seems I have been struck by a case of the raging clichés), 'if you always do what we've always done, you always get what you always got!'

Maybe I am being unfair, maybe your operation is already steady and predictable and you are ready for the challenge of becoming a lean enterprise.

Either way, the tools and techniques in this book, especially changeover time reduction, offer you a clear direction to follow.

The one trait, I have observed, common to all successful people whether they are billionaires or team leaders, is that they all had a plan. The good news is that you have already taken the first and most important step to introduce lean. You have recognised that you need guidance. You decided to buy this book. I don't claim any credit for the technical content; the subjects have been covered many times by more qualified people. The one thing which I can claim is that I have reworked the subject matter into a form that can be easily understood. If you disagree, please let me know and I will endeavour to improve it. My email address is at the front of the book.

To help you get most from the topics covered, I have summarised some NUGGETS on the next page.

It only leaves me to say, enjoy your lean journey as you begin to introduce flow and pull into your operation. On your first attempt it might go 'pair-shaped' and the doomsayers in your business will delight and say, 'it'll never work, let's go back to the old ways!' But with a bit of luck, and a bit of preparation, and a bit of practice, it will work first time. (Remember, Proper Preparation Prevents Poor Performance.) Like everything in this life, the more you practice the luckier you get. Trust me, I'm an engineer.

David Sykes
September 2020

CREATING PULL AND FLOW – NUGGETS

- Stop thinking in terms of capacity; start to think in terms of THROUGHPUT when balancing production operations.
- Stop thinking parts per hour, start thinking of THE TIME to produce each part
- Understand that the customer sets the rate of demand by TAKT TIME
- Plan to produce parts to the DRUMBEAT of the customer
- Balance activity to Takt time between production processes, especially assembly operations
- Use STANDARD WORK COMBINATION TABLES to capture activities and identify WASTE
- Reduce changeover time dramatically by first separating EXTERNAL from INTERNAL activities
- EXTERNALISE as much of the changeover as possible

- Introduce FLOW where possible by creating dedicated lines or CELLS
- Where you cannot FLOW, introduce PULL using KANBANS
- Minimise downtime and move to zero defects by introducing JIDOKA and POKA-YOKE
- Ensure your processes are STABLE and control them using SPC
- Ensure greatest flexibility by training operators in a range of skills. Record them on a SKILLS MATRIX
- Create a CURRENT STATE MAP for your process
- Use the 7-Tools of Quality to reduce WASTE
- Look to create FLOW between operations
- Where you cannot create flow use PULL and control it with PRODUCTION KANBANS and WITHDRAWAL KANBANS from a SUPERMARKET
- Create the FINAL STATE MAP
- Implement the FINAL STAGE MAP or you have just wasted your time!

APPENDIX ONE

CREATING STABILITY – A RÉSUMÉ

At the end of the first book I offered the reader these 'nuggets' on how to create stability in a business:

- Identify the **seven wastes** in your business
- Make sure you have the right **metrics** in place
- Focus on **productivity** or standard labour hours recovered
- When confronting the challenges of your enterprise, spend time **defining the problem.** To brush over or overlook this step is to waste vital effort
- Use **the Seven Tools of Quality** to determine how to reduce them
- Do not overlook the value of **flowcharting** in this process
- Once the solution is identified, capture it as a **Standard Operating Procedure** (SOP)
- Organise the workplace using **5S**
- Remember the most important step, **Sustain**
- Use **Visual Factory** and **Visual Control** to simplify factory layout and report progress
- Introduce **Andon** and identify the **Natural Workgroup**
- Introduce **TPM**
- Work carefully with service departments to introduce **autonomous maintenance**

Let us consider briefly each of these:

WASTE

Waste is anything other than the minimum amount of equipment, materials, parts, space and employee's time which are **absolutely essential** to add value to the product or service. These are:

- **Overproduction,**
- **Waiting,**
- **Transportation,**
- **Over-processing,**
- **Inventory,**
- **Scrap/Rework**
- **Motion.**

METRICS

In this section we introduced Management Control and Reporting Systems (MCRS) and focussed on the terms Efficiency and Effectiveness

- A process is considered to be EFFICIENT when:
- "A given quantity of OUTPUTS cannot be produced with any less INPUTS"
- The "Rate of EFFICIENCY is simple the amount of (or value of) OUTPUTS divided by the amount of (or value of) INPUTS.

PRODUCTIVITY

Productivity (or Labour Efficiency) is defined as:

Productivity = <u>Man-Hours Recovered at Standard</u> x 100%
Total manned hours paid

PROBLEM SOLVING

Defining a problem

- There are many definitions, but in business terms, let us define a problem as:
- *'A gap between a perceived state and a desired state'*

THE SEVEN TOOLS OF QUALITY

Once we have decided what the problem is, the next stage of the process is to collect data so we can analyse it. This is where the Seven Tools of Quality come in, which I summarise below.

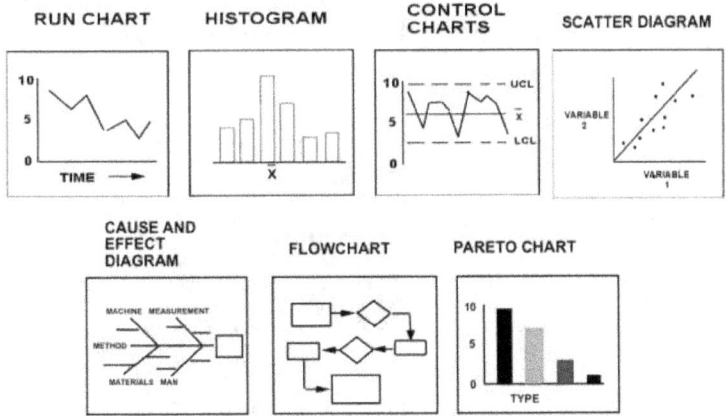

Run Chart

A run chart is where we collect data and plot it on a chart. Often this is the value of a variable we wish to control as the y-axis (vertical line) with time as the x-axis (horizontal line.)

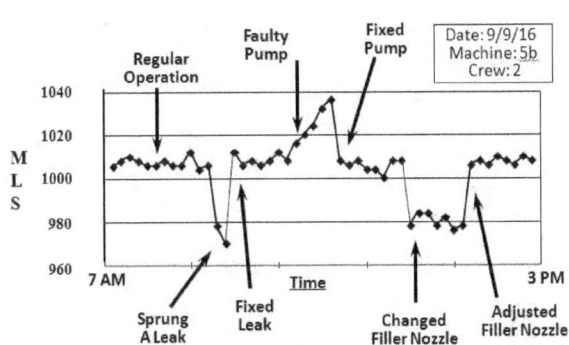

Filling 1L bottles of a shampoo

Histogram

If we were to plot the above data as a histogram or bar chart, the pattern would not be as clear. A Histogram is an effective way to organise data. Instead of looking at a table or list of points, we make a picture that shows how the points are "distributed" (i.e. where they are located on the scale)

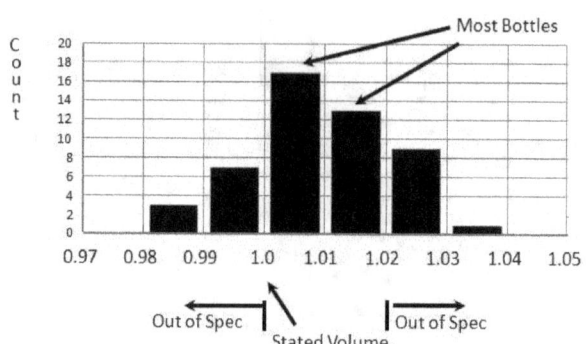

Filling 1 liter bottles of shampoo(Spec = 1.0 - 1.02 l.)

A **Control chart** is a tool for decision-making. It helps distinguish between common and special cause variation and uses statistically derived limits based on data from the process.

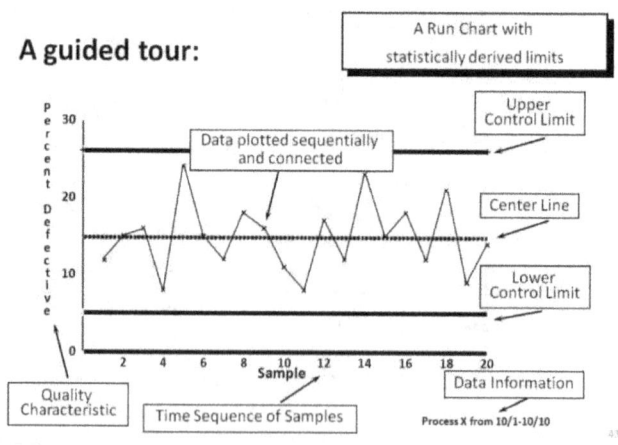

Scatter Diagram

A Scatter Plot shows the relationship between two variables. For example, the chart below shows that as children get taller, they weigh more. Each point on the plot represents an individual child for which a pair of measurements (height and weight) is available. The question is: "Are the height and weight-related?"

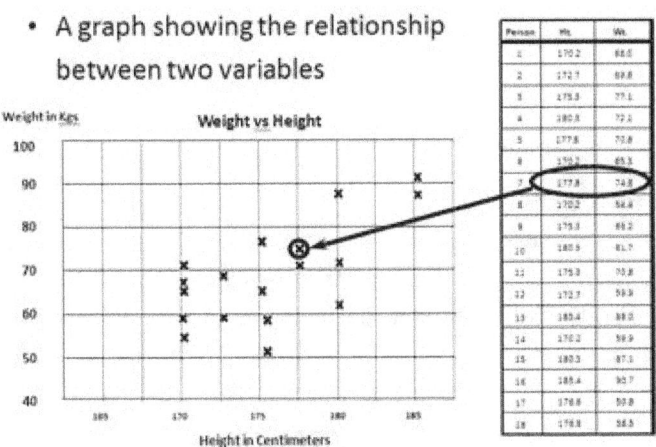

- A graph showing the relationship between two variables

A Run Chart is a special case of a Scatter Plot where the horizontal axis is time. In a Run Chart, the question is "How does a variable (on the vertical axis) change over time (on the horizontal axis)?"

Cause and Effect Diagram.

Often called a Fishbone Chart because of its shape, this is a very useful tool for getting to the bottom of a problem and helps us focus on the areas that require action, including collecting data. Whilst it can be used by an individual, it is best used as a brainstorming technique in a problem-solving team. To be successful, there should be little discussion about individual issues as they are listed. The idea is to capture, on paper or a whiteboard, all the elements that may contribute to the non-compliance or 'problem'.

163

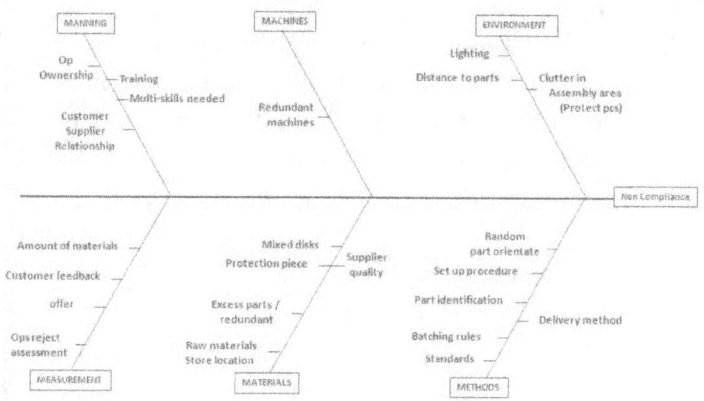

Flowcharts

A flowchart is a diagram that describes a sequence of operations. By breaking the process into its individual operations, the waste can be identified. Symbols are used to represent the operations and these are shown below:

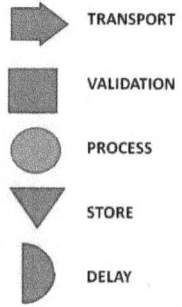

TRANSPORT

VALIDATION

PROCESS

STORE

DELAY

Pareto Chart

A Pareto Chart is a statistical technique used in decision making. It stems from what is known as the Pareto Principle. Pareto was an Italian economist who noted that 20% of the people controlled 80% of the wealth. This is the origin of the 80/20 rule you may have heard of.

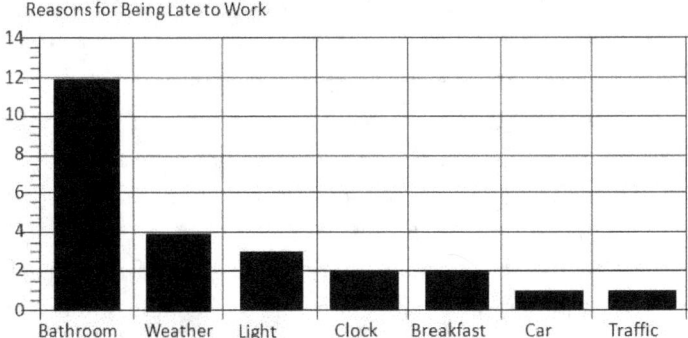

Reasons for Being Late to Work

STANDARD OPERATIONS

Henry Ford, the father of mass production techniques once said, "Give a lazy man an easy job and he'll find an easier way of doing it!" No doubt the lazy man was breaking the task down into its component parts and taking out the wasted steps, arrived at an easier way of working. Once this method has been identified, no doubt it would have been incorporated into an SOP. Our definition of a Standard Operation is:

A Standard Operation is centred around human movements, outlining efficient, safe working methods that eliminate waste whilst ensuring proper use of equipment and tooling.

Standard operations are used for training of staff, as a means of standardisation as well as a platform for improvements. They can also be used to identify and record safe working practices and are a tool to help manage the work-place. They act as an audit document to check that a person is actually performing the operation in the safest and most efficient way.

THE FIVE PILLARS - 5 S

The five pillars of 5S are:

Sort, Set-in-Order, Shine, Standardise and Sustain.

SORT means that we remove all items from the workplace that are not needed for the current production.

SET in ORDER is the logical outcome of the SORT exercise. Once the items that need to remain in the work area have been agreed, the next step is to arrange these in a way that means they are easy to use and easy to find and put away.

SHINE means sweeping floors, wiping down machinery and generally making sure that everything in the factory stays clean.

Once we have prepared the workplace, the next important element is to maintain it through **STANDARDIZATION.**

Perhaps the most difficult of the five pillars, **SUSTAIN** means making a habit of properly maintaining correct procedures. The four pillars can be only be implemented successfully if employees commit to SUSTAINING 5S

VISUAL FACTORY

Visual factory or visual management is the term used to represent a combination of signs, charts and other visual representation of information that enables the quick dissemination of data in a lean manufacturing process. It allows quick communication of information about the equipment's current operating conditions and environment through the use of Visual Aids.

It points out what is correct and identifies what has changed. This results in the quick identification of problems. It identifies when equipment is operating correctly and when not, instigating early treatment prior to failure.

ANDON

In traditional companies, the chain of command comes from the top and goes straight down. Lean operates with a different approach in that it is the process itself that initiates actions. One method is the use of Andon, a lean technique that empowers the operator to highlight abnormalities in the process and obtain support from above to continue to meet the plan.

An Andon system forms an important business measure of the health of a production line. It could be considered similar to the nervous system in our own bodies. By highlighting the concern in a timely fashion it maximises the chance of solving the problem before it affects productivity.

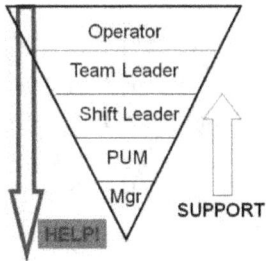

Andon helps support the principle of an inverse management pyramid where, instead of the workers being there to support the management and follow their instructions and requests, management is there to support workers and provide them with the help they need to resolve issues.

It is an information tool which provides an instant, visible and audible warning to the operations team that there is an abnormality within that area.

The key words are:

      INSTANT
      VISIBLE and AUDIBLE
      WARNING

Andon on it's on its own does not actually achieve anything. To fulfil its purpose, it needs a timely RESPONSE to make the Andon system effective.

NATURAL WORKGROUPS

Andon can only be effective if the team leader is backed up by an issue resolution procedure.

As mentioned earlier, the abnormality still needs to be addressed if it is not to affect the performance of that work area. This is done by identifying the 'natural workgroup' that interfaces with that work area. It is important that individuals are identified and are accessible to the workgroup. This means giving all operators the NAMES and NUMBERS of the members of the group.

Once an issue is raised, then the 'natural workgroup' must support that request as soon as possible. If the issue isn't eliminated in a timely fashion then the system should escalate to the next appropriate level. By doing so, the abnormality will eventually be addressed and a long-term countermeasure put in place to prevent future occurrence.

For example, earlier we referred to Airport security. If nothing happened when the alarm went off, it would be worthless. Ironically, whenever a car alarm goes off, instead of spurring people in the vicinity to action, most people will ignore it and more likely curse the car owner for disturbing them!

TOTAL PREVENTATIVE MAINTENANCE

What is TPM?

First and foremost, it is a tried and tested way of eliminating waste, saving money and making factories better places to work. As its name suggests, it is a system of maintenance covering the entire life of the equipment and the total human resource associated with it. No longer just the preserve of time-served craftsmen and engineers, it involves everyone in the maintenance of the production process. In doing so, it gives operators the knowledge and confidence to manage their own machines.

TPM is a long-term process which ultimately leads to increased skills, higher efficiency and zero losses. It can be the foundation for improvement for an entire production process. What it is not is simply repairing equipment as quickly and efficiently as possible when it breakdowns. Conversely, its aim is to minimise the chances of equipment ever breaking down, ensuring it runs as effectively as possible and, when it fails, repairing it as quickly and efficiently as possible.

AUTONOMOUS MAINTENANCE

This is an operator skills development programme which allows problems to be identified and solved quickly. It is an approach which, when implemented, stops the accelerated deterioration of the plant and equipment and deterioration related failures. It uses what we have already learnt in terms of SOPs through the development of standards. These standards ultimately result in training materials on how to run operate and maintain the equipment.

Autonomous maintenance is the primary way in which production departments can play a part in TPM. Operators are systematically trained in a step-by-step process for cleaning, checking and making minor adjustments to the equipment they work with. In doing so they bring operators and service departments together for a common goal.

Through educating and developing the workforce in autonomous maintenance, we can improve the operation and availability of plant and equipment. This is essential in achieving the first stage of stability required for our lean journey.

As a caveat I should add that introducing autonomous maintenance needs careful thought. There is a saying,

'If you don't train them, don't blame them'

If we are going to up-skill operators, it is important to involve those currently doing the work. Not to do so will cause resentment and this will frustrate the end goal of implementing autonomous maintenance. How to do so is outside the scope of this book but we will consider this in Book 3 when we discuss Leading4Lean.

By the Same Author

Leadership,
A Formula for Success

Do you often think you've got the right people but not the right result?

Instead of changing people, learn to 'change' your people by changing your leadership style. In this book you will learn...

- A Formula for Leadership as profound as $E = MC^2$
- The Power of Behaviour
- How your team can become Winners and Heroes
- Leadership is as simple as A-B-C
- How to harness the Vanilla Effect and watch motivation soar
- The 5 key steps to build trust and rapport
- The Assertiveness Toolbox
- How to supercharge your team for success
- Ten proven steps to successful Leadership

The book is based on a ten week leadership training programme which reached the finals of the National Training Awards and has delivered measureable, lasting results since 1997.

David Sykes is an experienced manager and trainer. A graduate chemical engineer, he has spent the majority of his career in man-management. Throughout he has shared his knowledge and experience of building and motivating successful teams.

In 2005, he founded Vanilla Training Solutions Ltd, a training consultancy dedicated to helping organisations excel. Having strong inter-personal skills he has coached and mentored managers and team leaders, particularly in their early careers.

ISBN 978-1-326-60865-1

Reviews

***** Real practical advice to improve your leadership skills, delivered in an entertaining and informative manner., 17 Jun. 2016
Andrew Wilson

Let me start by saying that I had the pleasure of working alongside David a few years ago. My abiding memory of our time together was the continuous and sometimes hilarious stories that he told regarding human behaviour. All the stories were pertinent and interesting and always somehow related back to a book or a management theory. I learnt over the year that we worked together that David was an ambassador of leadership, such was his passion for the subject.

Leadership is an enormous subject and one that I find fascinating and frustratingly hard to grasp such is the breadth of knowledge and opinion on the subject. What David has achieved in this book is singularly impressive. As an author, I know how hard it is to tackle a large topic without being totally consumed by it and paralysed to a point that you do not know where to start. I have found in my own experience that only through truly knowing your subject can you navigate around it with sufficient dexterity to write a book worth reading. In Leadership - a formula for success, David has accomplished something I have not seen in any other leadership text I have read in my 25 years of management; he has managed to cover a veritable smorgasbord of theory with a light-fingered dexterity only a master of the subject can do, and he has managed to do it in an entertaining and informative way. Many of the topics in this book are, on their own, heavy reading in the source texts, but David has managed to make them relevant by showing how they operate in the context of other management theories on the subject and brought them to life with real-life examples.

If you are looking for a source text on the subject of leadership, this book should form a foundation block in your reading library. You should not underestimate what you can learn from this book, it is easy to read and entertaining, and that is its secret; you will glide through the pages absorbing page on page of good sound tried and tested advice. You will learn things that you can put into practice immediately that will improve your leadership skills.

I recommend this book without hesitation.

***** Maslow meets Red Dwarf!, 12 Jun. 2016
Paul Hughes MSc FCIL
Any leadership book that combines 'real life' experiences coupled with academic analogies and Red Dwarf is a definite hit! You'll read it, make notes all over it and use it!

***** As a reader you feel like you are in the same room having a conversation ..., 23 Jun. 2016
John Aizlewood
David has captured the chemistry of people leadership in this very 'easy to read' book. It is packed with real-life examples that you can relate to and analyse. As a reader, you feel like you are in the same room having a conversation and it gets you thinking how you can deal with a similar situation, with the benefit of David's hindsight. Don't just read it, use it. Great practical tests as you go along that embed the theory set out in distinct chapters. After reading it you feel motivated to become a better leader. So, what are you waiting for? Click "add to basket" now!

***** I loved this book....27 July 2016
Amazon Customer
I loved this book and having completed it 4 weeks ago I can now see a difference in myself and how I behave at work. I like the approach the book took where at times it felt more like a conversation than a textbook and the practical examples give good context to the theory being discussed and its relevance for your daily working life. The module approach worked for me and I have been encouraging my colleagues and anyone who will listen how they need to give the book a chance and they won't be disappointed

***** What a great read....3 September 2016
Nicolas Nixon, Supply Chain Director, Coca-Cola Enterprises
What a great read. If you are a front-line manager or middle manager, this is an excellent book. Practical examples, anecdotes and models to help your leadership style. It is a book to get you thinking. Congratulations on a fantastic piece of work.

***** It's Brilliant....27th November 2017
Jason Davenhill, Performance Coach, Inflow Performance Ltd
C.H. pointed me in the direction of your book, Leadership, a Formula for Success. I can honestly say it is the best book on management and people stuff I have read. It's brilliant.

An Engineer's Guide to Influencing and Persuading

People do business with people they know, like & trust...

Without authority over people, if we are to achieve the results demanded of us by ourselves and our organisation, we have to have POWER WITH PEOPLE. Regardless of their purpose, all business is ultimately people business and understanding people is the key to our success.   Only by learning how to influence and persuade, therefore, can we get what we want. In this book you will learn:

- ■ The Power of Behaviour
- ■ The five key steps to building trust and rapport
- ■ Influencing is as simple as A-B-C
- ■ The Assertiveness Toolbox
- ■ How to motivate people to help you
- ■ The Lure of WIIFT (What's In It For Them?)
- ■ How to choose the right words
- ■ The Five-Step Plan for Influencing

If you want to know more about people, what drives them and motivates them and how you can use this knowledge to get what you want, this is the book for you.

David Sykes is an experienced manager and trainer. A graduate chemical engineer, he has spent the majority of his career in man-management. Throughout he has shared his knowledge and experience of building and motivating successful teams.

In 2005, he founded Vanilla Training Solutions Ltd, a training consultancy dedicated to helping organisations excel. Having strong inter-personal skills he has coached and mentored managers and team leaders, particularly in their early careers.

ISBN 978-1-326-82281-1

90000

Lean Made Simple – Creating Stability

Lean Made Simple
Creating Stability

The biggest barrier to using lean successfully is people's mindsets

If you've driven on a Smart motorway and enjoyed steady, hassle-free progress rather than the stop-start of old, you are experiencing lean in action. Instead of the traditional battle for space, the techniques employed change our collective mindsets allowing us to 'go with the flow.'

The principles, first brought into play in the car industry in Japan in the 1950's, are now an integral part of our everyday lives. Despite this, there is still a mystery perpetuated about Lean Thinking. This book explains the first vital stage in introducing lean thinking to your enterprise.

It covers the key subjects of Value and Waste, MCRS, the Seven Tools of Quality, 5S, TPM and Visual Factory in an easy to read way.

David Sykes is an experienced manager and trainer. A graduate chemical engineer, he has spent the majority of his career in manufacturing. Since 1988 when he became a JIT supplier to a major pharmaceuticals company, he has introduced lean principles in the businesses he has managed.

ISBN 978-0-244-06695-6

In 2005, he founded Vanilla Training Solutions Ltd, a training consultancy dedicated to helping organisations excel. He has worked as a lean consultant to a number of businesses.

www.ingramcontent.com/pod-product-compliance
Lightning Source LLC
Chambersburg PA
CBHW070338220526

45467CB00001B/163